ORIGINS

ORIGINS

OUR PLACE IN
HUBBLE'S UNIVERSE

———————————

JOHN GRIBBIN & SIMON GOODWIN

THE OVERLOOK PRESS
WOODSTOCK & NEW YORK

First published in 1998 by
The Overlook Press, Peter Mayer Publishers, Inc.
Lewis Hollow Road
Woodstock, New York 12498

Library of Congress Cataloging-in-Publication Data

Gribbin, John, R.
Origins: our place in Hubble's universe / John Gribbin, Simon Goodwin
p. cm.
Includes index.
I. Astronomy.   2. Astronomy—Pictorial works. 3. Hubble , Edwin
Powell, 1899-1953. 4. Hubble Space Telescope (Spacecraft)
5. Astronomers—United States—Biography
I. Goodwin, Simon, 1971-. II. Title.
QB44.2.G758   1997   520—dc21   97-21833 CIP

Originally published in Great Britain by Constable and Company Limited

Manufactured in the United States of America

ISBN 0-87951-813-8

1 3 5 7 9 8 6 4 2

This book is dedicated to the memory of Roger Tayler (1929–1997). Among many better-known and more important achievements, at either end of his time as Professor of Astronomy at the University of Sussex, from 1967 to 1997, Roger helped to set each of us on the path towards a career in astronomy.

CONTENTS

ACKNOWLEDGEMENTS

Thanks go to Bernd Aschenbach, Jennifer Ash-Poole, Ed Bell, Dave Leisawitz, Gary Linford, Jim Peebles, Jim Sahli, Steve Snowden and Peter Thomas for pictures and help with finding the right ones and also to Ian King and Jeremy Maris for help with image processing. Thanks for information to Anne Green, Paul Roche and Andrew Liddle.

AUTHORS' NOTE

This book follows the standard scientific practice in using the (American) billion to mean a thousand million (1,000,000,000) not the (British) billion of a million million (1,000,000,000,000). At the risk of irritating our astronomical colleagues, however, we have given distances in light years, the units favoured by science fiction writers, rather than the units usually used by astronomers, parsecs. A light year is the distance that light travels in one year, 9.46 thousand billion kilometres, and using this unit gives at least a hint of just how big the distances we are dealing with are (a parsec, by the way, is 3.2616 light years).

INTRODUCTION

Some seventy years ago, the American astronomer Edwin Hubble discovered that the Universe is expanding. The dramatic implication of this discovery, which was not fully appreciated for decades, is that, since the Universe is getting bigger, not only was it smaller in the past, but if you go back far enough into the past it must have had no size at all. It must have been born at a definite moment in time, in the Big Bang.

The reason why this discovery was not made sooner was that before the 1920s the telescopes available to astronomers were not adequate to reveal the true scale of the Universe. Hubble made his discovery with the aid of new technology – specifically, the 100-inch (2.5-metre) diameter reflecting telescope on Mount Wilson in California, which began operating in 1918 and was for almost three decades the largest and most powerful telescope on Earth. This instrument – still in use today – is now known as the Hooker Telescope, after the benefactor who provided the funds for its construction.

Hubble himself was quite a character. Born in Marsfield Missouri, on 20 November 1889, he was the fifth of seven children of a lawyer. He studied law himself, graduating from the University of Chicago in 1911, and visiting the University of Oxford as a Rhodes Scholar. During his time at Oxford, he represented the University at boxing, and fought an exhibition bout as an amateur against the French champion Georges Carpentier. He had been offered a chance to turn professional and fight the great champion Jack Johnson, but (surely wisely) turned it down. Returning to the United States in 1913, Hubble briefly practised as a lawyer, but soon decided that this was not the career for him, and went back to the University of Chicago to study astronomy. He received his PhD in 1917, and was immediately offered a job at the Mount Wilson Observatory; but the United States had just entered the First World War, and Hubble volunteered for the infantry. He went off to France to fight, where he was wounded by shell fragments in his right arm. So it was in 1919, at the age of thirty, that he actually began working on Mount Wilson, with the brand-new 100-inch telescope.

Before the Hooker Telescope became available, it seemed that the Universe consisted of an island of stars, the Milky Way, floating in a great, dark void. It is an impressive island – modern observations suggest that it is a disk-shaped system, about 100,000 light years in diameter (which means that light literally takes 100,000 years to cross the disk from one

(Opposite) Star trails. As the Earth rotates, the stars in the sky seem to revolve round the poles during the night. With a long-exposure photograph it is possible to catch this apparent motion, as in this stunning picture. The backdrop includes the dome of the Anglo-Australian Telescope at Siding Spring, Australia.

side to the other), made up of several hundred billion stars, each more or less the same as our Sun. But it is not alone. In the early 1920s, using the Hooker Telescope, Hubble established that what appear as fuzzy patches of light in lesser telescopes are, in fact, other islands in space far beyond the Milky Way, but comparable in size (some bigger, some smaller) to the Milky Way itself. The Milky Way system became known as our Galaxy, with a capital G; these other islands in space are called galaxies, with a small g. It is estimated that about fifty billion galaxies are in principle visible to the best modern telescopes, although only a few thousand have yet been studied in any kind of detail. Among other things, that means that there are a minimum of ten thousand billion billion stars (a 1 followed by 22 zeroes) in the visible Universe.

Of course, very few of these stars can be distinguished individually. Most galaxies are so far away from us that even the best modern telescopes only show the vast majority of them as fuzzy patches of light. With the Hooker Telescope, Hubble could resolve individual stars in just a handful of the nearer galaxies. But this was a major breakthrough. Because he could identify individual stars in the nearer of the neighbours to the Milky Way, he could measure their distances. There is a family of stars, known as the Cepheids,

Originally built in 1918, the 100-inch Hooker Telescope at Mount Wilson Observatory in the USA was the most powerful telescope in the world for nearly thirty years and is still in use today.

which vary in a regular way, brightening and dimming in a repeating pattern. The average brightness of any one of these stars determines the time it takes to run through one cycle of brightening and dimming. So, by measuring the length of the cycle for a particular star, astronomers can work out how bright the star really is; then, by measuring how faint it looks they can tell how far away it is. This is one of the most important techniques in the astronomers' toolkit, and it show us, directly, that even relatively nearby galaxies lie

(Opposite) Edwin Hubble at the 100-inch Hooker Telescope, with which he did much of his most important work.

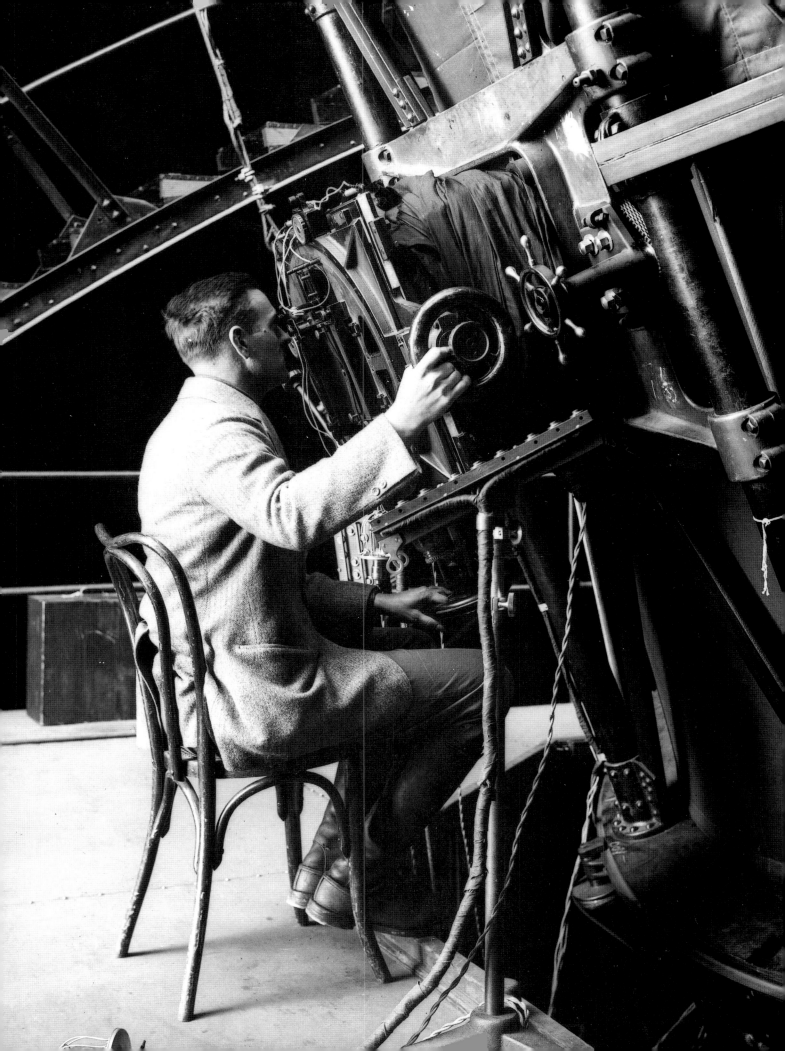

millions of light years beyond the Milky Way. We see those galaxies by light which left them before the genus *Homo* had even evolved.

This emphasises a dramatic feature of the way we see the Universe. Because it takes light a finite time to travel across space, we see objects further and further away as they were longer and longer ago. Instead of talking about the distance to a remote object astronomers sometimes refer to the 'look-back time'. A telescope is a kind of time machine, which we can use to get images of what the Universe was like when it was younger simply by looking further out into space. The snag, of course, is that the more distant an object is, the fainter it looks to us, and the more sensitive the detectors used to study the object have to be. But modern instruments are very sensitive indeed. Light travels at a breath-taking 300,000 kilometres per second, but even so there are objects visible to modern instruments which are so remote that we see them as they were not millions, but *billions* of years ago – more of this later.

If Hubble had retired in 1925, he would still be remembered for making one of the most dramatic discoveries in all of science, the existence of galaxies beyond the Milky Way. But he didn't stop there. Over the next few years, working with his colleagues at Mount Wilson, Hubble discovered that all but the very nearest galaxies to us seem to be receding from the Milky Way, and he quickly realised that in fact all the galaxies are receding uniformly from one another, except in cases where near neighbours are held together by gravity in a bound system (similar to the way the planets are held in orbit around the Sun by gravity, and the way all the stars in the Milky Way are held together in the disk-shaped galaxy by gravity). Hubble had discovered that the Universe is expanding, and he did so using one of the most famous, but also one of the most often misrepresented, tools of astronomy – the redshift.

When light from a hot object, such as a star or a galaxy, is split up into the rainbow pattern of the spectrum (perhaps by using a prism), the spectrum is seen to be crossed by distinct, sharply defined lines. Each element produces its own distinctive 'fingerprint' pattern of lines in the spectrum, always at the same wavelengths when studied in laboratories here on Earth. It is the presence of these fingerprints that enables astronomers to work out, among other things, which elements are present in the Sun and stars. But Hubble noticed that the patterns of lines in the spectra of light from other galaxies are all shifted towards the red end of the spectrum, by different amounts. (The colours of the spectrum are, in order, red, orange, yellow, green, blue, indigo and violet; red corresponds to longer wavelengths and violet to shorter wavelengths.)

One way in which features in a spectrum can be shifted bodily in this way is by the Doppler effect, named after the nineteenth-century Austrian physicist Christian Doppler, who predicted it in 1842. It works both for light and for sound. If an object is moving towards you and emits a musical note, the sound waves get squashed together by the motion and you hear a higher note than the one being played. Similarly, if the object is moving away, you hear a deeper note. This version of the Doppler effect is familiar in

(Opposite) The Hubble Space Telescope as it floats away from the space shuttle *Discovery* after the February 1997 upgrading mission.

everyday life from the way the note of the siren on an ambulance seems to deepen as the vehicle rushes past you. The optical version of the Doppler effect shifts lines in the spectrum towards the blue end of the spectrum if the object (for example, a star) is moving towards you, and towards the red end of the spectrum if the object is moving away. These blueshifts and redshifts are indeed seen in the light from stars, and at first Hubble interpreted the redshifts of galaxies in the same way – as indicating that they were rushing apart through space, like fragments of a bomb blasted outwards by its explosion.

But what Hubble didn't know when he discovered the cosmological redshift was that the expansion of the Universe had actually been predicted, ten years earlier, by Albert Einstein. And Einstein's prediction, based on his then brand-new general theory of relativity, said that the expansion was not caused by objects moving apart through space (remember, Einstein's equations made this prediction before the nature of external galaxies had been discovered), but by space itself expanding as time passed.

When Einstein found that the equations of the general theory had this expansion built into them, he was baffled. In 1917, it was thought that the Milky Way was the entire Universe, and the Milky Way certainly is not expanding. So Einstein added another factor to his equations, a new term, the cosmological constant, dropped in simply to hold the Universe still. He later described this as the biggest blunder of his career. What the combination of Hubble's observations and Einstein's theory (without the fiddle factor!) actually made clear at the end of the 1920s was that the Universe is indeed expanding, and that galaxies are being carried along for the ride as the space between them stretches, stretching the wavelengths of lines in the spectra of those galaxies as the light travels through expanding space on its way to us. The cosmological redshift is *not* a Doppler effect, and it does *not* mean that the Big Bang was an explosion involving a lump of matter sitting somewhere in the expanse of empty space; space itself began expanding in the Big Bang, and at no time was there ever anything 'outside' the explosion.

A useful way to picture what is going on is to imagine a fat piece of elastic, or a rubber band, on which you make marks with a biro. If you stretch the piece of elastic, every spot you have marked gets further away from every other spot; but none of the spots is moving through the elastic. The elastic corresponds to empty space, and the way that space stretches is described by the general theory of relativity.

Although the redshifts in the light from distant galaxies are not caused by the Doppler effect, they can still be thought of as due to motion of some kind. So they are usually measured in terms of velocity – a galaxy with a redshift of 100 km/sec, for example, is receding from us at that speed, because the space between us and that galaxy is expanding at that speed. But how far away would such a galaxy be? The crowning achievement of Hubble's career came at the end of the 1920s, when a series of painstaking measurements of both the distances (using primarily the Cepheid technique) and the redshifts of nearby galaxies showed that the redshift of a galaxy is proportional to its distance. In other words, a galaxy twice as far away from us is receding twice as fast. This is exactly the kind of expansion predicted by the general theory of relativity, and it is the only kind of expansion that would look exactly the same whichever galaxy you happen to be living in.

There is nothing special about the Milky Way, and observers in any other galaxy will also see the same pattern, of universal expansion with redshift proportional to distance. All of this was discovered by Edwin Hubble, which is why it is appropriate to describe the Universe we live in as Hubble's Universe.

The key to the Universe was the measurement of the relationship between redshift and distance. Once that crucial number, now known as Hubble's constant, was determined, the distance (and look-back time) to any galaxy could be determined simply by measuring its redshift. Measuring redshifts is relatively easy; the difficult part of the job was

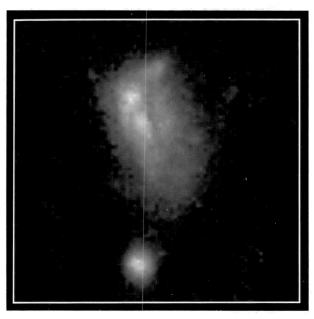

This galaxy is several billion light years away, which means we are seeing it as it was several billion years ago.

measuring distances to enough galaxies accurately enough to determine the value of Hubble's constant. There is still some uncertainty in the exact value of this number, but the uncertainty is getting smaller every year, as new observations are made. We have been involved (together with Martin Hendry, of Glasgow University) in one attempt to measure the Hubble constant, appropriately using data from the Hubble Space Telescope, also named in honour of Edwin Hubble. The calculations we made resulted, in 1997, in a figure of 55 km/sec/Megaparsec (close to the figure other scientists have come up with using different techniques), which means that a galaxy 1 Megaparsec (about 3.25 million light years) away is receding at 55 km/sec, while a galaxy 2 Megaparsecs away is receding at 110 km/sec, and so on. In fact, of course, astronomers use this relationship the other way round; they measure the redshift, and use that to work out how far away a galaxy is.

This doesn't just give us a distance scale for the Universe – it also tells us its age. The Hubble constant tells us how fast the Universe is expanding, so it also tells us how long it has taken to expand to the present size since the Big Bang. The smaller the value of the

Hubble constant, the older the Universe must be, because it would have taken longer to expand this much. For a Hubble constant of 55 km/sec/Megaparsec, the Universe cannot be more than 18 billion years old. In fact, Hubble's constant is not really constant, because the rate at which the Universe is expanding must have slowed down as time passed, because gravity is trying to pull everything back together; so it was expanding more rapidly when it was younger, and won't have needed quite so much time to reach its present size. Once allowance is made for this, the true age of the Universe must be between 12 billion and 15 billion years. In case you have been misled by some of the more sensational (and less accurate) newspaper accounts of these investigations, we should perhaps mention that this does indeed mean, as you would expect, that the Universe is older than the oldest known stars in the Universe. We shall use a figure of 15 billion years as a rough estimate for the age of the Universe.

Some 15 billion years ago, everything we can see in the Universe (all those fifty billion or so galaxies) was packed into one superdense, superhot fireball, the Big Bang. If you imagine winding the present expansion backwards to its most extreme beginning, it would imply that the Universe appeared out of a mathematical point, a singularity. Nobody believes that that is really what happened, and a new understanding of physics will be needed to explain what happened at the very beginning of time (the implications, beyond the scope of the present book, are discussed by John Gribbin in *In The Beginning*). But if we set that hypothetical moment as time zero, where (or when) can we begin to apply standard physics, the kind that has been tried and tested in experiments in laboratories and using 'atom smashing' machines here on Earth? If you make the calculation of how the density of the Universe has declined as it has expanded, then wind that calculation backwards mathematically, you find that just one ten-thousandth of a second after time zero everything in the visible Universe today was packed together in a sphere about one-sixth of a light year across, at the density of the nucleus of an atom (one hundred thousand billion times the density of water) and at a temperature of one thousand billion degrees C. The standard model of the Big Bang tells us, at least in outline, about everything that has happened since – the origins of stars and galaxies, planets, and even people. It tells us about our place in Hubble's Universe.

Until the 1960s, this picture of the Big Bang was not taken entirely seriously. Cosmologists made the calculations, and compared them with observations, and got the right answers. But somehow, nobody really felt in their bones that the calculations were giving a picture of the real Universe. There were only about a dozen cosmologists around, anyway, and they treated the calculations as a kind of game, like cosmic chess, an intellectual exercise rather than an investigation of where we came from. Their eyes were opened to the reality of what the equations were telling them by the discovery of the cosmic microwave background radiation, the echo of the Big Bang itself.

This radiation had actually been predicted back in the 1940s, but the prediction had been forgotten. With the aid of his students, the Ukrainian-born American George Gamow (an ebullient character who was possibly the only cosmologist of his generation to really believe that the equations did indeed describe the birth of the Universe) had calculated what would have happened to the energy of the primeval fireball as the Universe expanded and cooled. The calculation is relatively simple. It says that by about 300,000 years after time zero the entire Universe would have been filled with radiation (essentially, light) at a temperature of a few thousand degrees, roughly the same temperature as the surface of the Sun today. As the Universe expanded, this radiation would have cooled down (one way of looking at this is to think of the radiation that filled the Universe being redshifted to longer and longer wavelengths as the Universe expands). Gamow and

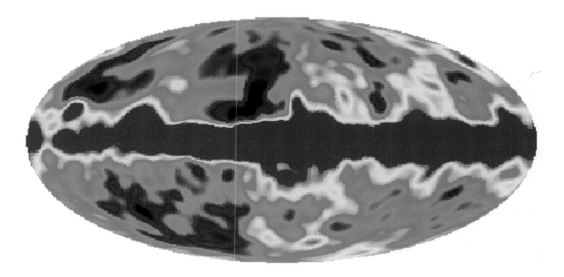

COBE all-sky map of the microwave background radiation. The thick red band across the centre shows the position of the Milky Way.

his colleagues calculated, some fifty years ago, that by now the Universe would still be filled with radiation, but that the radiation would have been cooled (redshifted) so much that it would have a temperature of about *minus* 270 degrees C, and would be in the form of cool microwaves, the kind of radiation that would be produced by a very feeble radar transmitter, or by an extremely cold microwave oven.

In the early 1960s, two radio astronomers, Arno Penzias and Robert Wilson, were working with a new antenna at the Bell Laboratories in New Jersey. This instrument had been designed and used in early experiments with satellite communications, and the astronomers were refurbishing it and setting it up to use for radio astronomy. Penzias and Wilson were baffled to find their new telescope plagued by a kind of interference, a weak hiss of radio noise, coming from all directions in space, with a temperature of −270 degrees. They had never heard of Gamow's prediction, but when their discovery was announced in 1965 the connection soon became clear. It was the discovery of the cosmic microwave background radiation that forced what few cosmologists there were to take seriously the idea that there really was an early Universe, and which encouraged many physicists to climb aboard the cosmological bandwagon. The combination of theory and observations fits together so beautifully that it simply cannot be ignored, and must be revealing a deep truth about the nature of the Universe we live in.

One of the greatest achievements of human thought has been the development, over the past seven decades (but especially over the past three decades), of this understanding of how the Universe evolved from a superhot, superdense state – the Big Bang – into the state we see it in today. Using the known laws of physics, studied in laboratories here on Earth, and comparing these with the predictions of Albert Einstein's general theory of relativity, astronomers can calculate how an expanding Universe filled with hot gas cooled, and how that gas condensed to form stars and galaxies, planets, and, ultimately, ourselves. For many years, the observations used to back up these theories consisted largely of fuzzy photographs and squiggles produced by electronic detectors, largely incomprehensible to the lay person, and downright dull as images. But now all that has changed. The latest generation of detectors produces beautiful images from deep space that show the process described by these theories at work, and which stand out as stunning works of art, regardless of their scientific importance.

The archetypal example of such a detector is the Hubble Space Telescope, which has provided so many glorious pictures of the Universe that they formed the subject of a

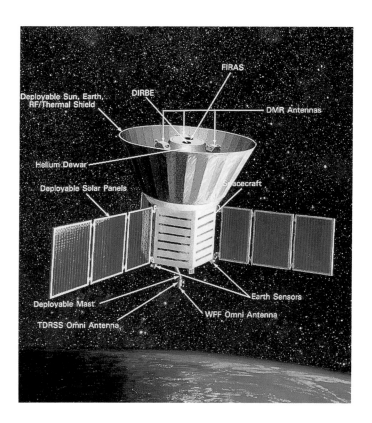

An impression of the Cosmic Background Explorer satellite in orbit. COBE was launched in 1989 and the mission finished after four years, when the coolant required for two of the three instruments finally ran out.

previous book (*Hubble's Universe* by Simon Goodwin). But there are other equally impressive images, and we have gathered some of them together here. Some come from satellites orbiting above the Earth's atmosphere – satellites such as COBE, the COsmic Background Explorer, which provided an image of the afterglow of the Big Bang, a view of the Universe almost unimaginably far back and far away – just 300,000 years after time zero. Space probes have sent back images closer to home, of the planets and other objects in our own Solar System, which shed new light on the processes by which the planets (including the Earth) formed. And ground-based telescopes, continually being upgraded with improved detectors, have provided images in the middle ground – of galaxies and stars being born. Many of the images we have used to tell this part of the story come from the Anglo-Australian Telescope, a 3.9-metre diameter reflector sited 1,150 metres above sea level on Siding Spring Mountain in Australia.

But we have not forgotten the Hubble Space Telescope itself (the HST), which continues to provide some of the most spectacular views of deep space. Launched in April 1990,

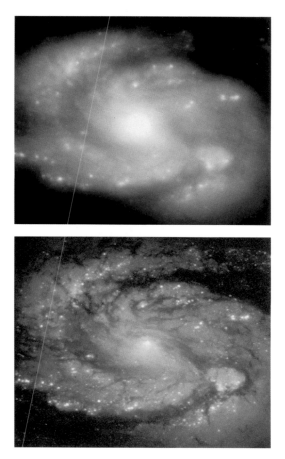

Images showing the vast improvement in the Hubble Space Telescope's power after the 1993 repair mission, when a new instrument was added to the satellite to correct the flaw in the main mirror. The upper picture shows a blurred image of the spiral galaxy M100, before the repair, and the lower one shows the spectacular results that could be achieved after the repair.

the HST initially suffered severely from a kind of astigmatism, caused by its main mirror, 2.4 metres across (roughly the same size as the Hooker Telescope that Hubble himself worked with) having been incorrectly manufactured. It was only after a repair mission at the end of 1993 that, as was explained in *Hubble's Universe*, the telescope was able to function properly, producing the pictures of deep space that made a whole generation of non-astronomers aware of the new understanding of the Universe. In February 1997, a routine servicing mission visited the HST once again, replacing some of its old instrumentation and adding new instruments, including the Near Infrared Camera and Multi-Object Spectrometer (NICMOS), to probe the Universe at wavelengths longer than those of the red part of the spectrum, beyond the range visible to human eyes.

To give you some idea of the value of this kind of upgrade, remember that the original instrumentation for the HST was based on designs from the early 1980s. The upgrade is like replacing an original IBM PC (first put on the market in 1981) with the latest state-of-the-art Power Macintosh. These improvements should keep the HST itself functioning as a state-of-the-art machine until the end of the twentieth century, with the prospect of further upgrading on further servicing missions in 1999 and at later dates – one further mission is already scheduled for the year 2002, and it is hoped to keep the HST operating until 2005.

Already, though, astronomers are looking even further into the next millennium. Outline plans have been drawn up for several different kinds of Next Generation Space Telescope (NGST), one of which may well be launched early in the twenty-first century. It is the success of the HST that has convinced the holders of the purse strings that this will be a good investment for the taxpayer's money – an estimated $500 million to build and launch the NGST, and $400 million more to operate it for a decade. Other probes planned for the new millennium include a mission originally known as COBRAS-SAMBA (from COsmic Background Radiation Anisotropy Satellite/SAtellite to Measure Background Anisotropies), but recently given the simpler but less informative name Planck Explorer. This satellite will be launched around the year 2005, and will map the cosmic microwave background radiation in unprecedented detail. And there are still new telescopes being built on the ground. With the help of lighter materials for their construction, and better computers to steer them and analyse their observations, these are larger and more powerful than Hubble could have dreamed. For example, the Keck Telescope on Mauna Kea, in Hawaii, has a mirror 10 metres in diameter, made up from thirty-six separate hexagonal mirrors, each 1.8 metres across, which fit together in an array like a slice through a honeycomb, and are steered by computers to work together as if they were one big mirror.

Not that you always need such sophisticated equipment to produce a pretty picture of a heavenly object. Astronomy is just about the last area of science where amateurs can, and do, still make important contributions, and just for fun we include here a picture of Comet Hyakutake taken with an ordinary 35-mm camera using a 50-millimetre lens – but, admittedly, a photograph taken from an observatory in the clear air of a mountain top on Tenerife.

Comet Hyakutake. This photograph shows that you don't need hundreds of millions of dollars' worth of satellite to obtain a good astronomical picture. It was taken by Luis Chinarro at the Observatorio del Teide in Tenerife with a 50-mm lens on a normal camera, using normal film and a 10-minute exposure time.

By putting images from different sources together, and with the aid of the phenomenon of the look-back time, we are able to tell a coherent story, providing an illustrated history of the Universe from the Big Bang to the present day. Using an array of telescopes and detectors, some based on the ground and others orbiting in space, astronomers have been able to obtain images of the Universe in all its stages of development, providing the definitive answer to the age-old question of how we come to be here at all.

That story is spelled out in detail, step by step, in the images and accompanying text which form the heart of this book. But in order to set the scene, we'd first like to give you an overview of the modern understanding of how a sea of superhot gas, expanding out of the Big Bang and cooling as it did so, has given rise to galaxies, stars, planets and people.

The story begins one hundred-thousandth of a second after time zero, when conditions in the entire Universe resembled the activity that goes on in particle accelerators today when beams of particles (such as protons) are smashed head on into one another. Because those conditions can be studied, if only in a limited way, in experiments today, physicists are confident that they know what went on in the early stages of the Big Bang, and how particles such as protons, neutrons and electrons (the building blocks of ordinary atomic matter) interacted at that time. Their calculations say that, by the time these interactions stopped, and the Universe had cooled to the point where stable atomic nuclei could exist (about four minutes after time zero), three-quarters of the atomic matter in the Universe should have been in the form of hydrogen (the simplest element) and one-quarter should have been in the form of helium (the next simplest element). Lo and behold, when astronomers study the composition of the oldest stars, which formed soonest after the Big Bang, they find that they are indeed made up of 25 per cent helium and 75 per cent hydrogen.

But stars did not begin to form until millions of years after time zero. Structure in the Universe grew out of tiny irregularities that were left over from the Big Bang, ripples which left their imprint on the cosmic microwave background radiation. Regions that were slightly more dense than the neighbouring regions had a stronger gravitational pull, because they contained more matter, and this attracted more matter, so that the irregularities grew as the Universe expanded. Huge sheets of gas, hydrogen and helium left over from the Big Bang, began to collapse under the influence of gravity, and broke up to make vast collections of galaxies, forming walls and sheets which stretch far across the Universe. Within those galaxies, smaller clouds of gas collapsed under the influence of gravity to form stars. Heavier elements formed planets.

But this could not have happened if the bright stars and galaxies that we can see were all there is to the Universe. When astronomers try to model the way structure grows in the expanding Universe, using computer simulations, they find that the gravity of all the bright stars and galaxies is not sufficient to do the job. There must be a lot more dark matter, tugging on the bright stuff gravitationally, and holding everything together even though the Universe is expanding. Since this dark stuff cannot be seen, it is not very photogenic, so we cannot show you pictures of it. But the illustrations on pages 33 and 35 show you how the pattern of galaxies in the real Universe compares with the kind of pattern that the computer modellers come up with if they allow for the presence of a hundred times as much dark matter as there is bright stuff. The visible Universe is actually much less than the tip of the proverbial iceberg.

Within a galaxy like our own Milky Way, things would have been pretty dull if all the visible matter had stayed in the form of hydrogen and helium. But the way in which a star generates energy, holding itself up against the inward tug of gravity, is by the process known as nuclear fusion. In the first stage of this nuclear burning, hydrogen is converted into helium, and energy is released. This process is going on inside the Sun today, stopping it from collapsing and keeping it shining. At later stages in the life of a star, helium may be converted into the heavier carbon, carbon into oxygen, and so on. All of the elements in the Universe (except for the original hydrogen and helium) have been manufactured in this way. At the end of their lives, when their nuclear fuel is exhausted, the bigger stars explode, scattering the elements built up inside them (plus many more produced in the explosion itself) across space. Those heavy elements are incorporated into clouds of gas and dust in space, from which new stars, laced with heavier elements, can form. Among other things, planets are made out of these heavier elements, and planets – such as Earth, with its iron core – could not exist until the first stars had run through their life cycle in this way.

That is how the Solar System formed. The Sun is a middle-aged star, only some 4.5 billion years old, and formed out of material that had already been partially processed inside two or more generations of preceding stars. Everything in the Solar System, except hydrogen and primordial helium, including the atoms your own body is made of, has been manufactured in this way inside another star.

The origin of our Solar System is intimately related to the structure of the Milky Way galaxy, just as the origin of our Galaxy is intimately related to the structure of the whole Universe. The Galaxy is, as we have mentioned, a disk-shaped system of stars. Seen from above, it would look something like a huge Catherine wheel (or like the galaxy shown on page 57), with tightly wrapped 'spiral arms' of stars winding round the central bulge. The Sun and its family of planets lie about two-thirds of the way out from the centre, towards the rim. Unlike a Catherine wheel, a galaxy like our own does not rotate as a solid body. Individual stars and other objects move in their own orbits, and pass through the spiral

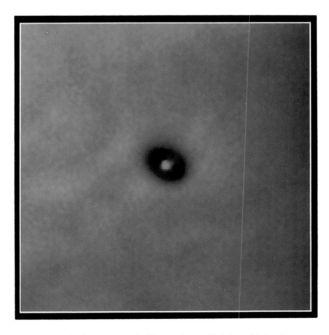

A star in the Orion Nebula surrounded by a dusty disk in which planets are forming.

arms on their journey round the centre. The spiral arms are features rather like waves on the ocean, through which stars pass all the time.

These spiral arms stand out because they are made up of hot, young stars which shine brightly. The kind of stars forming today – stars like the Sun – form in groups, loose clusters of stars all formed from one large collapsing cloud of gas or dust. And what makes them collapse is their passage through the spiral arms. It may just be the squeeze they get from the pressure wave associated with the arm that does the trick. It is more likely, though, that the trigger for a burst of star formation is the blast wave from a supernova, the explosion of a large star, rippling through the interstellar medium. Stars of all sizes form in the spiral arms, and the most massive stars live for only a few million years before they explode, seeding the nearby clouds with heavy elements, and also causing some of those clouds to collapse and form new stars. All of this happens within the vicinity of the arms, because these short-lived stars do not have time to move far before they explode.

Because of all the second-hand material around in the cloud from which the Sun formed, it was accompanied by a dusty disk of gas, in which particles of dust collided and stuck together and gradually grew big enough to attract other dust grains by gravity. The lumps grew bigger until they formed the planets, with the last stages of planet formation involving the bombardment of the young planets by huge rocks from space as the last of the debris was swept up. The scars from this terminal bombardment can still be seen today, notably on the faces of the Moon and Mercury. Some of the material left over from the formation of the Solar System formed a band of rocky debris (the asteroid belt) orbiting between Mars and Jupiter; some formed a cloud of objects orbiting beyond the orbits of the planets, from which one lump is occasionally disturbed into a trajectory that takes it close past the Sun and puts on a temporary, but spectacular, display as a comet.

One of the planets that formed from the swirling disk of dust around the young Sun was just the right distance from the Sun for oceans of liquid water to form and provide a home for life. But every step in the chain, from the Big Bang to life on Earth, depended on the previous step. We are what we are because the Universe is the way it is.

Stars like the Sun, though, live for much longer than the kind of star that forms a supernova, and the story of life on Earth is less than half over. Our Sun is roughly half-way through its lifetime, and has about another 5 billion years of existence in more or less its present state to look forward to. Travelling once around the galaxy every 250 million years, it has already completed the circuit about twenty times. When the Sun has exhausted its hydrogen fuel, after another twenty circuits of the Galaxy, it will spend a short time burning helium into carbon, and its atmosphere will swell up, engulfing the inner planets. Eventually, being too small to form a supernova, it will fade away into a stellar cinder, a cooling lump of star stuff about as big as the Earth.

For every step in this story, the match between theory and observation is good, confirming that astronomers really do understand the origins of the Universe and all it

(Opposite) The Horsehead Nebula: one of the most spectacular examples of a dark nebula seen against the backdrop of a bright emission nebula.

contains. The stunning new images which we have selected for this book show the birth-pangs of creation in the Universe at large at each step in the chain. We can now see pictures of events that previous generations of astronomers could only picture in their mind's eye – and which non-astronomers could not even imagine.

What we see confirms that we live in a Universe that was born in a Big Bang, and has been expanding for some 15 billion years, while stars, and planets and people came into existence within it. This is the Universe discovered by Hubble at the end of the 1920s – Hubble's Universe. Our place in Hubble's Universe is as the inhabitants of an ordinary planet, orbiting an ordinary star, in a backwater of a slightly smaller than average galaxy, one of tens of billions of galaxies in the Universe at large. There is nothing special about our place in the Universe – but the view is spectacular.

THE PICTURES

—— PLATE 1 ——

THE COBE FOUR-YEAR MAP

The most distant view we have across the Universe, and the furthest back in time that we can 'see' towards the Big Bang, is the view provided by the cosmic microwave background radiation. This is actually detected by radio telescopes, both on the ground and on board unmanned satellites such as COBE. But astronomers can use the measurements of the radio noise that fills the Universe to draw a kind of contour map of how the sky would look to us if our eyes were sensitive to this kind of microwave radiation. The colours they use to fill in these maps are in principle as arbitrary as the convention by which the old British Commonwealth countries are coloured pink on maps of the world; but the usual convention, as with the COBE image opposite, is to colour regions of the sky that are slightly hotter than average (corresponding to greater density) red, and regions of the sky that are slightly cooler than average (corresponding to lower density) blue.

This image was built up by adding together in a computer data obtained over the entire four-year lifetime of the COBE mission. It shows the entire sky, as two hemispheres, in the same way that some flat maps of the Earth show the entire surface of the planet as two hemispheres. The detail shows 'ripples' which reveal the nature of irregularities in the Universe when it was just 300,000 years old. The radiation from those ripples has been travelling through space for some 15 billion years, giving us a snapshot of how the hot gas that filled the Universe at that time was distributed. Those irregularities were so big that they formed the seeds from which not just individual galaxies but huge clusters and super-clusters of galaxies grew – the largest features seen by any optical telescope on Earth would fit within the smallest features seen on this map.

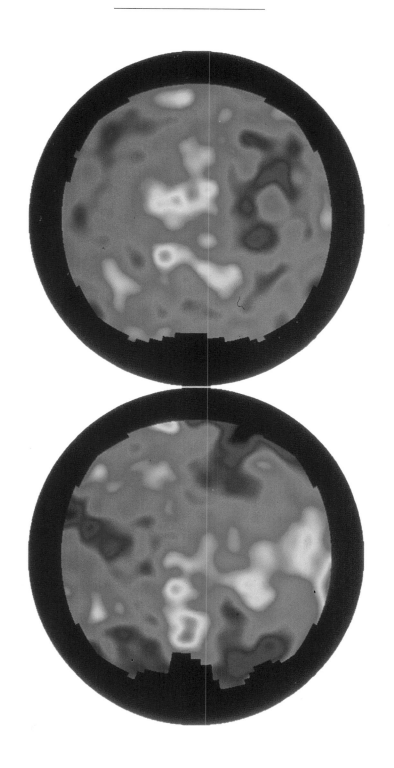

—— PLATE 2 ——

SIMULATIONS OF GALAXY FORMATION

For about a billion years after the time that the background radiation last interacted with matter, the Universe was dark. The first galaxies took time to grow out of the ripples imprinted on the expanding Universe as it emerged from the Big Bang. In order to fill the gap, and to test their theories of how galaxies formed, astronomers use large supercomputers to calculate the way in which ripples like the ones revealed by COBE must have grown. The computer program calculates how an initially almost imperceptible irregularity gets bigger by attracting surrounding material through its gravitational pull. It cannot use observations of the details of the actual ripples in the background radiation to calculate the formation of real clusters of galaxies because there is not enough detailed information in the observations of the background radiation. Instead, it is based on a pattern of initial ripples which has the same overall statistical properties as the ripples in the real Universe, but an arbitrary detailed structure.

This image is the only one in this book which does not come directly from the real Universe. It shows the result of this process, simulated in a Cray-T3D supercomputer (one of the most powerful computers in the world), which mimics the simultaneous collapse of individual lumps of matter and the expansion of the Universe which separates the lumps. The result is a complex, filamentary structure, with regions of high density stretching in streamers and sheets across the Universe, surrounding large empty voids. According to theory, this kind of structure should have been imprinted on the Universe by the time it lit up, as galaxies and stars started to shine in the densest regions.

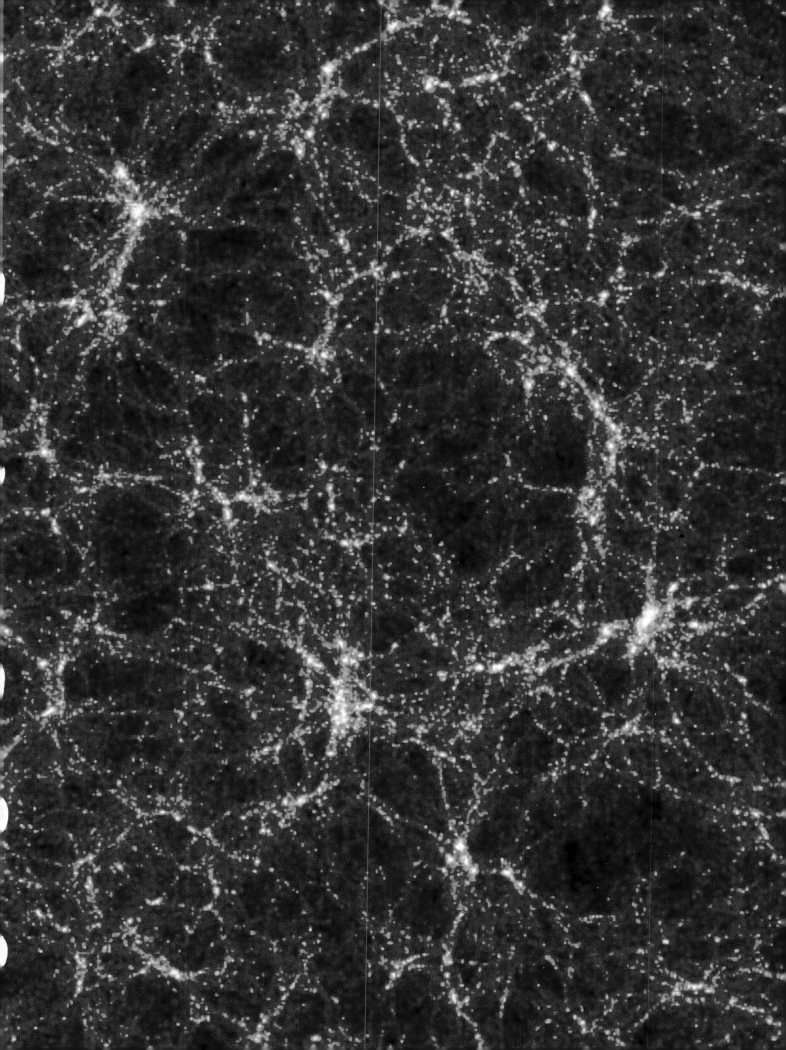

— PLATE 3 —

THE LICK GALAXY MAP

One of the most dramatic pieces of evidence that astronomers really do understand the nature of the Universe at large is provided by the match between computer simulations of the growth of structure in the early Universe and what we see on the night sky. By putting together a mosaic made up from many separate photographs taken at the Lick Observatory in California, astronomers at Princeton University compiled this view of the northern hemisphere night sky which contains the images of more than a million galaxies. The scale is so large that you cannot see all the individual galaxies; bright spots in this picture represent clusters of galaxies, which each contain hundreds or even thousands of individual galaxies. The striking bright spot near the middle of the picture, for example, is the Coma Cluster.

The most remarkable overall feature of this portrait of more than a million galaxies is the way in which its overall appearance resembles that of the computer-generated image on page 33. This is confirmation that the computer simulations really do provide a good link between the time when the Universe emerged from the hot fireball of the Big Bang and the time when it settled down into more or less the state we see it in today. To give you some idea of the scale involved, the largest dark voids in the visible Universe are about 250 million light years across. Because light travels at a finite speed, the furthest we can see across the Universe is 15 billion light years, the distance that light has had time to travel since the Big Bang. So an individual void is about one-sixtieth of the size of the observable Universe.

The Lick Galaxy Map

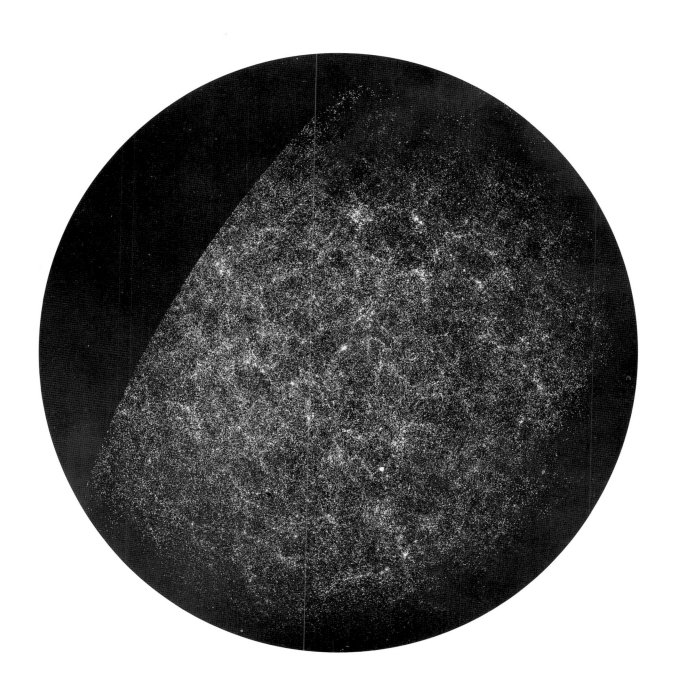

——— PLATE 4 ———

THE COMA GALAXY CLUSTER

Our own origins are intimately connected to the origin of structure in the Universe, because without those ripples in the fireball that grew to become great streamers and superclusters of galaxies there would be nothing in the Universe except a thin gruel of hydrogen and helium gas, getting thinner as the Universe expanded. Even our Milky Way Galaxy, let alone the Solar System, is a tiny detail in the overall structure of the Universe. Some idea of just how small we are on the cosmic scale comes when we take the first step down from the Universe at large to smaller structures; this brings us to clusters of galaxies like the Coma Cluster. But even the Coma Cluster is made up from more than a thousand bright galaxies, held together by gravity and moving together like a swarm of bees. It is more than 300 million light years away from us, and is being carried further away at a rate of 6,700 km per second by the expansion of the Universe.

One of the best ways to see that the Coma Cluster really is a single system, and not a chance alignment of many galaxies on the sky, is by looking at X-ray wavelengths. X-rays do not penetrate the Earth's atmosphere, and X-rays from space can only be studied by hoisting instruments on rockets and satellites such as ROSAT into space. ROSAT (short for ROentgen SATellite, named in honour of the discoverer of X-rays) was launched in 1990, and carried out a complete survey of the sky at low-energy X-ray wavelengths. As with radio images of the Universe, the X-ray data can be converted into colourful maps, which show how the world would look if we really did have Superman-like X-ray vision. Most of the X-rays from the Coma Cluster come from hot gas between the galaxies, although some individual galaxies show up as bright spots in the image. The concentration of gas towards the centre of the cluster shows that the whole system is held together by gravity, forming a kind of gravitational pot-hole, with the gas falling into the centre of the system. The slightly brighter blob in the lower right of this X-ray image shows where a smaller group of galaxies is being sucked in by the gravity of the Coma Cluster, falling into the pot-hole, where it will eventually merge indistinguishably into the cluster.

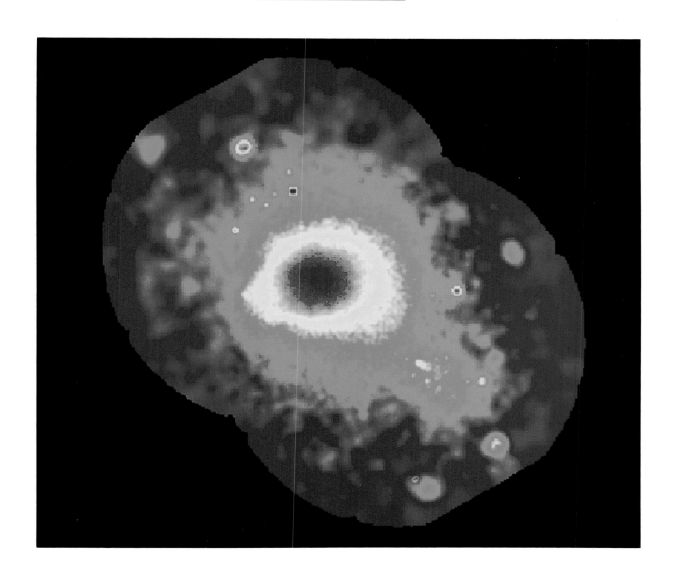

—— PLATE 5 ——

THE HOMES OF QUASARS

Even when we get down to the scale of individual galaxies, we find that the further back in time we look (that is, the further away across the Universe), the more violent the Universe was. The most extreme individual phenomena that we can see are the quasars. These each shine a thousand times more brightly than all of the hundreds of billions of stars in a galaxy put together, but the source of this brightness is energy from a tiny region, no bigger across than our Solar System. This is what gave them their name, a contraction of quasi-stellar object, because they look like points of light (like stars) in a telescopic image, but shine far more brightly than any star.

The explanation of this enormous energy output is that each quasar is a black hole, containing perhaps 100 million times as much mass as our Sun. Although this is only a small fraction of the total mass of a galaxy like our own, the concentration of so much mass in a small volume of space creates a gravitational pot-hole with sides so steep that nothing, not even light, can escape from it. In a young galaxy, there is plenty of gas around which has not yet been turned into stars and can fall into this black hole. As material is sucked into the black hole, it forms a swirling ring of matter around it, which is where the energy that powers a quasar is generated, and where sometimes whole stars are ripped apart and swallowed.

The most distant quasars can be seen 12 billion light years away across the Universe, as they were when the Universe was only 20 per cent of its present age, long before the Earth formed. But in spite of their vast energy output, such distant objects are too faint to be very photogenic. The ones pictured here are much closer to us, close enough that the host galaxies in which they are embedded can also be seen. Their distances range from 1.5 billion light years to 3 billion light years from Earth – so the light by which we see the closest one has been travelling for a third of the time the Earth has existed, and the light by which we see the most distant of these six systems has been travelling for two-thirds of the time the Earth has existed. The distorted structure of the galaxies is a result of the tidal forces of the quasars that are gripping them by gravity.

Even closer to home, many galaxies may harbour black holes which used to be quasars, but have gone quiet as the supply of gas and dust falling into them has run out; even our own Milky Way Galaxy has a supermassive black hole at its centre.

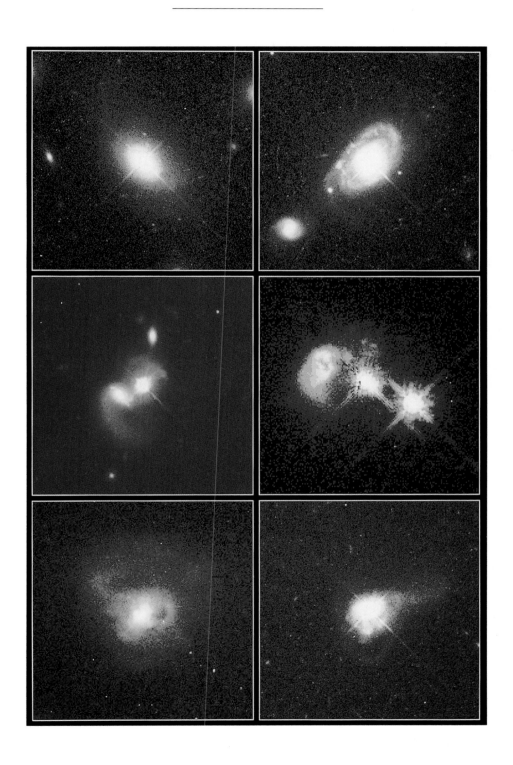

—— PLATE 6 ——

THE HUBBLE DEEP FIELD

The best view we have of very young galaxies in the Universe at large comes from the Hubble Space Telescope, which was used to take 342 exposures of the same patch of the sky, each lasting for between 15 and 40 minutes, on 150 successive orbits of the telescope around the Earth. These 342 exposures were then added together electronically to provide one image of some of the faintest (and most distant) objects ever viewed in the Universe. The Hubble Deep Field Survey covers a tiny portion of the sky, only one-thirtieth of the diameter of the full Moon as seen from Earth. The patch of sky chosen for the survey lies in the direction of the North Galactic Pole, pointing straight up out of the plane of the Milky Way to one of the darkest parts of the sky; but it is thought to be representative of the way the Universe would look in any direction, if the view were not obscured by the Milky Way itself. All of the hundreds of images in the Deep Field are galaxies, some so faint that they may be even more distant than the quasars shown on page 39. This means that the galaxies as we see them are very young objects – although the light by which we see them is now nearly as old as the Universe itself. What we see are swarms of irregular galaxies, not yet settled down into the smoothness of middle age, many of them in the process of interacting with other galaxies and merging to form larger systems.

—— PLATE 7 ——

A VERY YOUNG GALAXY

A close-up of part of the Hubble Deep Field image shows the way galaxies interacted in the crowded conditions that existed when the Universe was young. At distances corresponding to a look-back time of about 5 billion years (to a time just before the Earth formed), galaxies are much bluer than nearby galaxies, which is a sign that star formation was very active then (young stars are hot and blue). Many of the objects in these distant clusters are pairs of disk galaxies, broadly similar to the Milky Way, in the act of merging. Mergers like this produce a huge burst of star formation, using up most of the gas and dust in the merging galaxies. The merged systems then settle down into the form of smooth elliptical galaxies, which are largely free from gas and dust. About half of all the galaxies we see today have been involved in mergers with galaxies of a similar size within the past 8 billion years – roughly the second half of the lifetime of the Universe so far.

Disk galaxies themselves are thought to have formed from mergers between even smaller units very early in the life of the Universe. These small, faint objects are too distant to be seen in the act of merging, even with the HST. But the range of ages of the globular clusters in our own Galaxy (see page 60) is from about 14 billion years to about 8 billion years, suggesting that our Galaxy formed over a span of several billion years, from an amalgamation of about a million primordial gas clouds.

The very youngest galaxies, like the one marked by an arrow here, are so far away that the redshift has moved all the blue light from young stars into the red part of the spectrum. These galaxies, which look red to us, are further away than the furthest quasar. They are the most distant individual objects that have yet been seen.

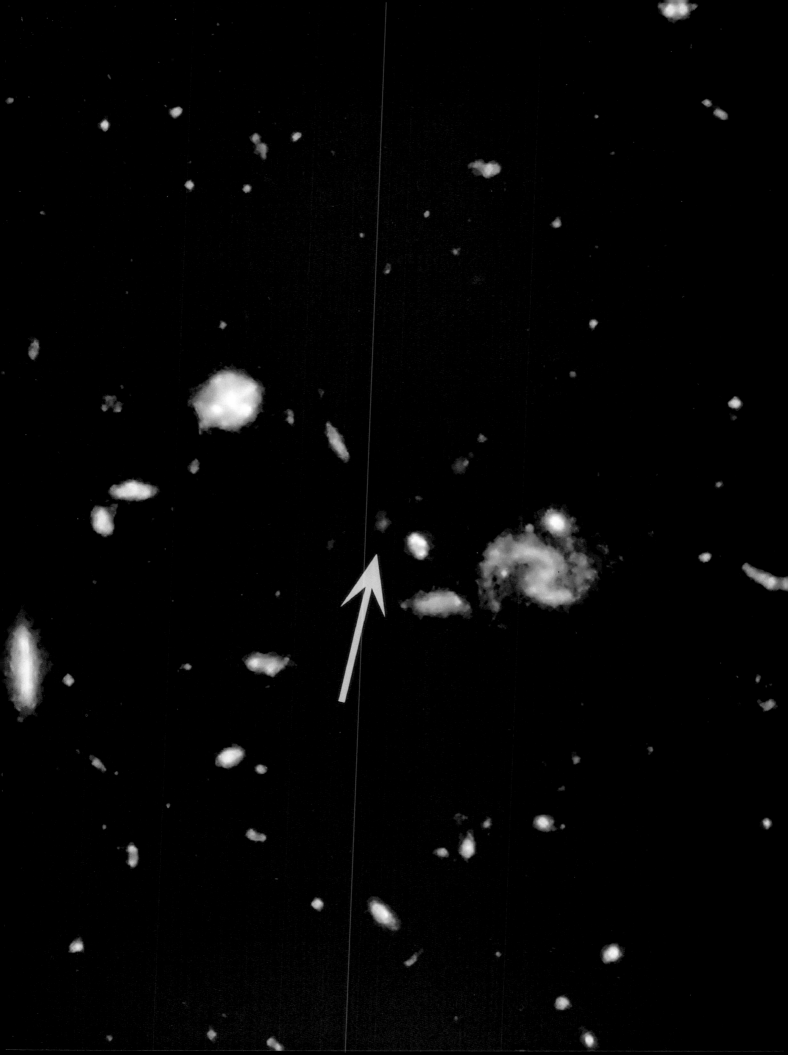

—— PLATE 8 ——

A GRAVITATIONAL LENS

Although very young galaxies are too distant to be imaged in detail by the HST without help, nature sometimes provides a natural enlarging lens which makes some of the detail inside one of these galaxies visible. If there is a large massive object along the line of sight between us and a very young galaxy, the gravity of the massive object can bend light from the even more distant galaxy, acting as a gravitational lens and magnifying the image. The effect was predicted by Albert Einstein, using his general theory of relativity, in the 1930s; but it was not observed until the 1980s.

In this image, obtained using the HST, light from a distant galaxy is being bent as it passes through the cluster of galaxies that can be seen near the centre of the image. The cluster is 5 billion light years away from us; the light from the distant galaxy, which is itself about 10 billion light years away, is bent to form five separate blue images, which have been distorted into arcs by the lensing effect. The blue images are near the centre of the picture, and at 6, 7, 8 and 2 o'clock.

The combination of the HST and the natural gravitational lens makes it possible to pick out details as small as 300 light years across in this galaxy 10 billion light years away. This is like being able to read lettering three-hundredths of a millimetre high at a distance of a kilometre. The detail shows that this is a lumpy galaxy in which primordial gas clouds are still merging to form a single system, with a great deal of star formation going on.

—— PLATE 9 ——

ELLIPTICAL GALAXY M87

Elliptical galaxies are the giants of the Universe, huge star systems built from the mergers of smaller galaxies. Although some objects described as dwarf ellipticals each contain only about a million times as much mass as our Sun, these resemble the globular clusters that are associated with our own Galaxy and others. The record-breakers are the giant ellipticals, each of which may contain several thousand billion stars. About 60 per cent of all galaxies are ellipticals, and they are particularly common in clusters of galaxies, where there has been ample opportunity for disk galaxies to collide with one another and merge to form ellipticals. In very rich clusters (ones in which very many galaxies are crowded together), there is almost invariably a huge elliptical galaxy at the centre of the system, where it sits like a spider at the centre of its web, dominating the cluster gravitationally and pulling more of the other galaxies in the cluster into its embrace.

This particular elliptical galaxy is M87 (also known as NGC 4486), one of the largest and brightest ellipticals in the relatively nearby Virgo Cluster of galaxies. Although it looks placid enough, sitting in space surrounded by its own globular clusters, in this picture from the Anglo-Australian Telescope (AAT), spectroscopic studies have shown that at the centre of M87 there is a whirling disk of matter, orbiting at a speed of 550 km per second around a central mass containing three billion times as much matter as our Sun. This can only be a black hole, and may once have been a quasar.

The red appearance of M87 is typical of elliptical galaxies, showing that they are made up of old, red stars. The presence of a central black hole is also probably typical; the most powerful astronomical radio sources are associated with giant ellipticals, and the most likely explanation is that the energy which powers the radio sources comes from matter falling into a black hole.

—— PLATE 10 ——

A FACE-ON SPIRAL GALAXY

The kind of galaxy we live in is much more photogenic than an elliptical, and has a lot more interesting activity going on even in its outer regions today. These are properly known as disk galaxies, because they are flattened, rotating systems; many disk galaxies, like NGC 2997 (photographed here by the AAT) have a beautiful spiral pattern when seen face on, and are also known as spiral galaxies. This is very much the way our Milky Way Galaxy would look if viewed from above; in our Galaxy, the Sun is about two-thirds of the way out from the centre, near one of the spiral arms.

The blue colour of NGC 2997 is typical of disk galaxies. It is caused by the presence of many hot, young, blue stars, which form along the spiral arms. All disk galaxies have a central bulge, dominated by older, red stars, like a miniature elliptical galaxy; the ratio of the disk to the bulge is roughly the same as the relationship between the white and the yolk of a fried egg, and the thickness of the disk is only about one-fifteenth of its diameter. Spectroscopic studies show that in galaxies like this the whole visible system is embedded in a cloud of dark matter (perhaps ten times as much as all the matter in the visible bright stars put together), holding the galaxy in its gravitational grip and controlling its rotation, which is always in the direction corresponding to winding up the spiral arms (clockwise in the image shown here).

—— PLATE 11 ——

AN EDGE-ON SPIRAL GALAXY

This is the same sort of galaxy as NGC 2997, a spiral galaxy, but seen edge on. The dusty clouds which provide the raw material for the birth of stars in disk galaxies show up very clearly in this AAT photograph of NGC 4945. Because the Sun and the Earth are located in the disk of our own Galaxy, this is similar to the view we get of the Milky Way from the inside – see page 65. About 30 per cent of all galaxies are disk galaxies.

—— PLATE 12 ——

A BARRED SPIRAL GALAXY

Many spiral galaxies, like NGC 1365 (pictured here by the AAT), have bars across their centres, with the spiral arms growing out from the ends of the bar. Computer simulations show that the spiral pattern seen in many disk galaxies would not persist for very long unless it were stabilised by the large halo of dark matter in which the galaxy is embedded. But even with this stabilising influence, the spiral pattern cannot persist for ever, and the natural way for the pattern to break up is by the growth of a bar of stars outwards from the centre of the galaxy.

It is likely that our own Galaxy will evolve in this way, and there is some evidence that there is already a small bar of stars across the centre of the Milky Way. But it is very difficult to observe this region of the Milky Way because the view is obscured by all the gas and dust in the disk of our Galaxy.

—— PLATE 13 ——

A STARBURST IRREGULAR GALAXY

Along with the easily classifiable elliptical and disk galaxies in the Universe, there are systems which are irregular in shape and do not fit either category – so they are known as irregular galaxies. About 10 per cent of all galaxies are irregulars, and many of them seem to be undergoing an intense phase of star formation, as in the case of this 'starburst irregular', NGC 1313, photographed by the AAT. Using radio telescopes, astronomers can often find evidence of a disk of gas within such an irregular galaxy. Together with the fact that stars are still being formed in them, this gives irregulars a family resemblance to the disk galaxies rather than to the ellipticals.

Starburst activity (which is also seen in otherwise normal disk galaxies) is often triggered by the tidal pull of a nearby galaxy interacting with the material of the starburst galaxy to produce dense regions of dust and gas which collapse to form stars; but in some cases starbursts seem to be occurring in galaxies that do not have any near neighbours, and we do not know what has triggered the activity (it may be that the galaxy has simply run into a cloud of cold gas between the galaxies, in 'empty' space). The large amounts of gas and dust associated with the star-formation activity make it difficult to see deep into these galaxies using ordinary visible light, but they show up very brightly to infrared detectors.

—— PLATE 14 ——

SPIRAL GALAXY M101

Because we cannot look at our own Galaxy from the outside, it is worth looking at another image which gives a different kind of view of a galaxy very similar to our Milky Way. This is M101, photographed in ultraviolet light by the Ultraviolet Imaging Telescope carried on the Astro-2 mission of the space shuttle *Endeavour*. M101 is only about 16 million light years away from us, making it a relatively near neighbour on the cosmic scale, and its appearance is typical of disk galaxies today.

Ultraviolet light lies beyond the blue end of the spectrum visible to our eyes – it is literally bluer than blue – and because hot young stars produce a lot of blue and ultraviolet light they show up particularly clearly in this kind of image, sharply outlining the spiral arms in M101. The image also highlights huge glowing regions known as HII regions – nebulae which shine because of the ultraviolet energy being produced by the young stars embedded within them. These are the birthplaces of stars.

SPIRAL GALAXY M101

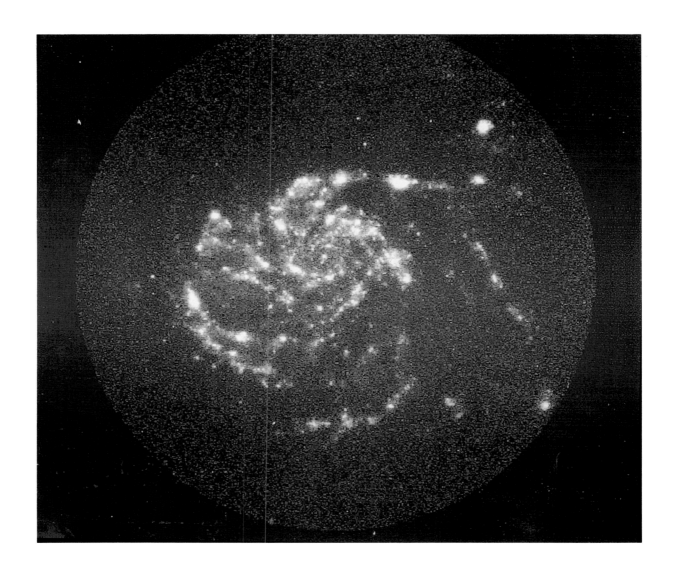

—— PLATE 15 ——

THE LARGE MAGELLANIC CLOUD

Focusing even closer to home, the same Astro-2 mission, carried out between 2 March and 18 March 1995, brought back this image of a star-forming region in the Large Magellanic Cloud. The LMC is an irregular galaxy which is a close neighbour of the Milky Way, at a distance of just 160,000 light years. It is so close that the light by which we see it left as recently as during the Ice Epoch before last here on Earth.

The bright region of star formation forms a clearly visible arc in this ultraviolet image. It is about a thousand light years long, and contains at least five clusters of very hot massive young stars. Because these stars are so massive, they use up their nuclear fuel quickly and run through their life cycles in only a few million years, many of them destined to explode and scatter heavy elements across the nearby region of space. But within these brightly visible regions there are also stars like the Sun, much smaller and fainter than the blue giants, but destined to have a much longer life, and still be around billions of years from now.

The shape of the arc of star formation may be an indication that it is part of a wave of star-forming activity rippling through the LMC. Such a wave can sustain itself because of the way massive stars quickly run through their life cycles and explode, the blast waves from the explosions squeezing clouds of gas and dust just ahead of the arc and triggering the next burst of star formation. A similar process probably explains the way in which spiral arms are formed in disk galaxies.

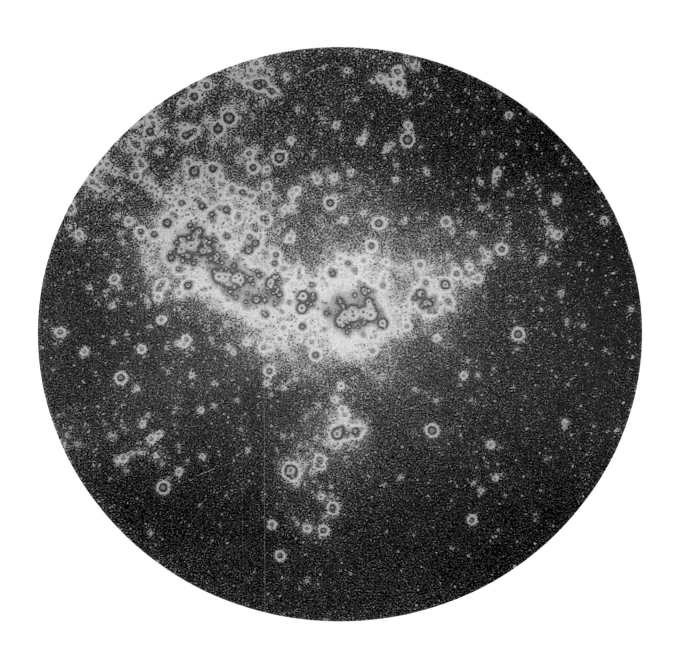

—— PLATE 16 ——

GLOBULAR CLUSTER 47 TUCANAE

This is a significant step in the story of our origins. For the first time, we have come close enough to home to give you a picture of an object within our own Galaxy. This is the globular cluster 47 Tucanae (also known as NGC 104), photographed by the AAT. It contains about a million stars in a spherically symmetrical ball, held together by gravity, and it lies in the bulge of our Galaxy (see page 65), about 30,000 light years away from us. It is a typical globular cluster, one of about 150 known to be associated with our Galaxy. At the heart of a globular cluster, the stars are so densely packed together that there may be as many as a thousand stars within a cube of space three light years along each side. To put this in perspective, if such a cube were centred on our Sun there would be no other stars within it.

The way that globular clusters are distributed, forming a sphere around the Milky Way, shows that they formed when the Galaxy was young, before most of its material had settled down into the disk of the Milky Way. Spectroscopy shows that the stars in globular clusters contain very little in the way of heavy elements, confirming that they formed out of almost primordial material, and 47 Tucanae itself is one of the older globular clusters associated with our Galaxy, with an age of between 12 and 14 billion years. The oldest globular clusters are therefore among the first stellar populations that formed after the Big Bang, and were incorporated into what became our Galaxy (and other galaxies) when the Galaxy began to grow and attract material to itself by gravity. In a sense, the oldest globular clusters are older than the Galaxy itself.

AN OPEN STAR CLUSTER

The Sun and Solar System did not form as part of a globular cluster, but from a cloud of gas and dust which collapsed and fragmented to form a more irregular open cluster of stars. The open cluster in this image (NGC 6520, photographed by the AAT) happens to lie near to the position on the sky of a dust cloud like the one from which the Sun and a handful of other stars formed. The cloud looks like a hole in the profusion of stars of the Milky Way, through which we are looking out into the depths of space; in reality, it is a cloud of dark material blocking out the light from the stars behind it. The cloud has nothing to do with NGC 6520; they just happen to be close enough together on the sky to appear in the same photograph.

An open cluster may contain anything from a few dozen stars to a thousand or so stars, in a region a few light years across. They contain hot, young stars that have recently formed in the disk of our Galaxy, but because there are so few of them they are not held together in a permanent group by gravity. As the stars move around the Galaxy in their own in-dependent orbits, they spread out, and the open cluster loses its identity. The stars that formed alongside our Sun some 4.5 billion years ago in an open cluster like this have long since gone their separate ways (the Sun has travelled right round the Galaxy about 200 times since then) and can no longer be identified.

—— PLATE 18 ——

THE MILKY WAY IN INFRARED

The 'fried egg' appearance of our own Milky Way Galaxy shows up clearly in this image, obtained by an instrument sensitive to infrared wavelengths of light, carried on the COBE satellite (see page 20). Because infrared radiation penetrates dust much more easily than visible light does, this instrument, the Diffuse Infrared Background Experiment (DIRBE), had a much clearer view of the stars in the disk and bulge of our Galaxy than the view available from telescopes operating in the part of the spectrum visible to our eyes. But because DIRBE operated in the near infrared, only just outside the visible band, most of the radiation it 'saw' did indeed come from stars, although these are not resolved individually and all we see is the overall pattern.

This image combines data from different parts of the sky obtained over a period of six months. It has been mapped onto a projection of the whole sky, as if the heavens were unwrapped and laid out flat; this is similar to the way in which the surface of the entire spherical Earth can be mapped onto a flat piece of paper using the Mercator projection. Like the Mercator projection, this map is distorted near the poles, at the top and bottom of the map – but in this case there is nothing much there to be distorted.

The middle of the picture is centred on the view towards the bulge of the Milky Way (the centre of our Galaxy), turn the page sideways and the left and right edges of the picture should be imagined as wrapped round to join behind the back of your head, giving the view out towards the edge of the disk. But because most of the radiation detected by DIRBE comes from the much denser population of stars towards the galactic centre, and very little comes from the thin population of stars in the opposite direction, there is an illusion of viewing the Milky Way from the outside (compare this with the image of the edge-on spiral NGC 4945 shown on page 51). The image certainly gives a feel for how far away we are from the centre of the Galaxy. As a bonus, some of the globular clusters surrounding the Milky Way can also be seen in this picture.

———— PLATE 19 ————

THE CENTRE OF THE MILKY WAY

You can get some idea of why the infrared detector on the COBE satellite was needed to obtain the image on page 65 from this view of part of the Milky Way in visible light, obtained using a wide-angle exposure on the Anglo-Australian Telescope. There are certainly plenty of stars to be seen; but the plane of the disk of our Galaxy is also full of dark clouds of gas and dust, which obscure the view towards the heart of the Milky Way – once again, it is worth comparing this image with the picture of NGC 4945 on page 51.

But although all of the dust in the plane of the Milky Way hinders astronomical observations of the galactic centre, we should not be too annoyed by its presence, since without it we would not be here. It is the dusty clouds in galaxies like our own Milky Way and NGC 4945 that are the birthplaces of stars and planets, including our Sun and Solar System.

—— PLATE 20 ——

THE CONE NEBULA

This is a stellar birthplace in our own Galaxy, the young star cluster NGC 2264. When a star forms from a contracting ball of gas, it gets hot in the middle and starts to glow. The heat comes from gravitational energy released by the collapse of the proto-star, and it produces a pressure which slows down the contraction of the ball of gas. But the amount of heat released is not enough to stop the proto-star collapsing, and it continues to shrink slowly and get hotter and hotter inside. When it gets hot enough, at a central temperature of about 1.5 million degrees C, nuclear reactions begin to take place in its centre, and these generate enough energy to stop the star from collapsing any more – at least, until the nuclear fuel runs out. A star like our Sun, burning nuclear fuel steadily and staying the same size, is said to be a member of the main sequence of stars. But the stars in NGC 2264 are only a few million years old, and are still contracting – they have not yet reached the main sequence where nuclear reactions begin to take place.

The striking dark streak in the bottom of the picture, a cloud of cool material silhouetted against the bright background, is known, for obvious reasons, as the Cone Nebula.

— PLATE 21 —

THE TARANTULA NEBULA

One of the most spectacular regions of star formation in our relatively near neighbour-hood can be seen in this AAT image of the Tarantula Nebula, 160,000 light years away in the Large Magellanic Cloud (not far from the site of Supernova 1987A; see page 81). It gets its name because of a fancied resemblance to the spider; more prosaically, it is part of a system known as the 30 Doradus complex.

The Tarantula is about a thousand light years across – so big that, even at the distance of the LMC, it covers half a degree of arc on the sky (about the same as the Moon) and can be seen with the naked eye. Another famous nebula that can be seen with the naked eye is the Orion Nebula, only about 1,300 light years away in the constellation of the same name, within our own Galaxy. The Tarantula system is thirty times the size of the Orion Nebula, but a hundred times further away; if the two were swapped the Tarantula Nebula would not only be visible by day, but would cast shadows at night.

The whole complex contains about 500,000 times as much mass as our Sun, and will probably form a globular cluster made up of at least 100,000 stars. Much of this mass is now in the form of very hot, very young and very massive stars at the heart of the nebula – it is estimated that there are about twenty stars, each with a mass between 100 and 200 times the mass of our Sun, as well as a profusion of smaller stars. The energy output from all the stars in the centre of the nebula is 50 million times the energy output of the Sun. A great deal of this energy is in the form of ultraviolet radiation, which is absorbed by the nebula and makes it glow (not unlike the way in which a UV light at a disco makes a white shirt glow).

The Tarantula Nebula is also visible in the ultraviolet image of the LMC on page 59, near the point where the bar across the image bends upwards.

—— PLATE 22 ——

THE LAGOON NEBULA

Although much more modest in size than the Tarantula Nebula, the Lagoon Nebula, shown here in an image obtained by the HST, is much closer to home, so it can be studied in detail. This system is very like the kind of stellar nursery in which our Solar System was born.

The Lagoon Nebula (also known as M8) lies 5,000 light years from Earth in the direction of the constellation Sagittarius. The light we see it by left the nebula about the time that the first stage of Stonehenge and the Pyramid of Giza were being built on Earth. The bright central region of the nebula (upper left in this image), called the Hourglass, is illuminated by the radiation from a hot central star. Together with other hot stars in the nebula, the energy output from this star is producing a strong stellar wind which is ripping apart the cool clouds of material around the stars.

Where the surfaces of the clouds are heated by the stellar wind, a blue mist of material is being driven off and blown away into space on the right-hand side of the picture. The difference in temperature between the hot surfaces of the clouds and their cold interiors, combined with the pressure of the stellar wind, has twisted some of the clouds into spiralling streamers, like interstellar tornadoes, each half a light year long.

The picture also shows a variety of smaller features within the nebula, including small dark blobs known as Bok globules. These are tiny clouds (by astronomical standards), each only about 8,000 times as big across as the distance from the Earth to the Sun, on the point of collapsing to form stars. Each globule contains somewhere between a tenth and ten or twenty times as much mass as our Sun.

—— PLATE 23 ——

A STARBURST IN NGC 253

If you were still under any illusions that the Earth and the Solar System occupy any kind of special place in the Universe, this image should shatter those illusions once and for all. It is not part of the Milky Way, but the central region of a starburst galaxy (see also page 55), known as NGC 253, which lies 8 million light years from Earth, in the direction of (but far beyond!) the constellation Sculptor. This HST image shows a region about 1,000 light years across, in which a huge number of stars are forming simultaneously. Each bright white blob in this picture represents not a single star, but a stellar nursery like the Lagoon Nebula, in which many stars are forming.

Galaxies like these are shrouded in dust (which is what the new stars and planets form from), so a great deal of the light from the young stars is obscured. This radiation (largely blue and ultraviolet light from hot young stars) is absorbed by the dusty clouds, and re-radiated in the form of infrared radiation, so starburst galaxies show up very brightly in infrared images of the sky. Many starburst galaxies were discovered by the Infrared Astronomical Satellite (IRAS), which flew in space in the early 1980s. Starburst galaxies are often otherwise ordinary disk galaxies involved in collisions or tidal interactions with other galaxies, which have triggered the burst of star formation.

There are more than 100 billion stars in our Galaxy alone, and more than 50 billion galaxies in the visible Universe, each roughly the same sort of size as our Galaxy, and there are millions more stars being made in some of those galaxies all the time. Even if the odds against any particular one of those star systems containing a planet like the Earth, with blue skies and running water, are very small, there are so many star systems out there that there must be millions of planets like the Earth somewhere in the Universe.

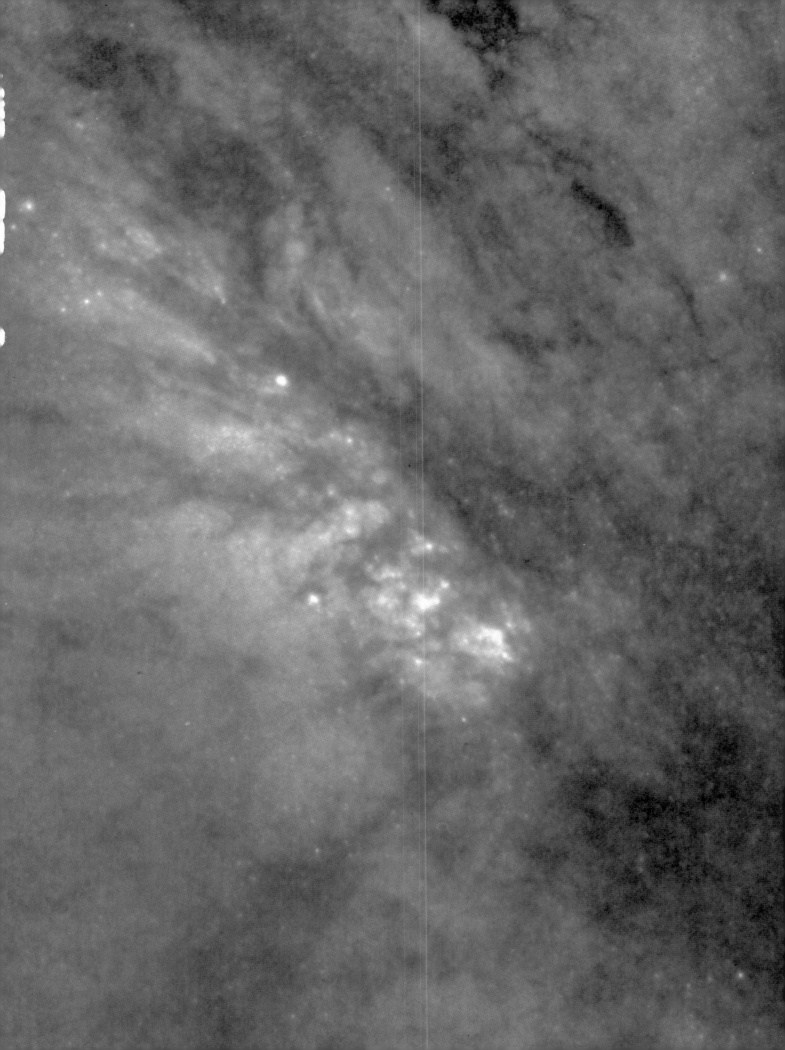

—— PLATE 24 ——

STARBIRTH IN M33

Even in ordinary galaxies which are not undergoing a starburst there are likely to be regions in which there is a great deal of star-forming activity. One of these regions, a nebula known as NGC 604, is pictured here in an image from the HST. NGC 604 lies in the outer regions of the disk of an ordinary spiral galaxy (very similar to the Milky Way) called M33. M33 is just 2.7 million light years away from us, in the direction of the constellation Triangulum.

There are at least 200 hot, young stars at the heart of NGC 604, each of them with a mass between twenty-five to sixty times the mass of our Sun, plus (it is assumed) many smaller stars. The energy from these stars is absorbed by the gas in the nebula, which glows as the energy it has absorbed is re-radiated. This image strikingly reveals the three-dimensional structure of the nebula, with cavernous holes in the cloud made by the pressure of the energy being radiated by the stars inside it, pushing the material away into space. Eventually, the cloud will disperse and the stars will settle down as an open cluster (see page 63).

— PLATE 25 —

COMETARY GLOBULE CG4

Looking more like a creature from a science fiction horror movie than anything else, this is in fact a gas cloud known as a cometary globule, and given the prosaic catalogue number CG4. Cometary globules have nothing at all to do with comets, but get their name from the superficial resemblance to a comet seen in the night sky from Earth. The head of such a globule is larger than the entire Solar System, and the tail (only part of which is visible in this AAT picture of CG4) stretches for more than 200 billion km – about 100,000 light minutes, or a fifth of a light year.

Cometary globules are a by-product of starbirth, in nebulas like NGC 604 (see page 77). They form when a region of hot gas being blown out from the vicinity of a hot young star (or stars) collides with cooler material in space. In this case, the collision between the hot, glowing material of the cometary globule and cold, invisible gas has given the head of the globule the appearance of a mouth about to devour a galaxy. But this is simply a trick of perspective: the edge-on disk galaxy to the left of the picture is actually millions of light years away, far outside our own Galaxy, while the cometary globule is part of the Milky Way.

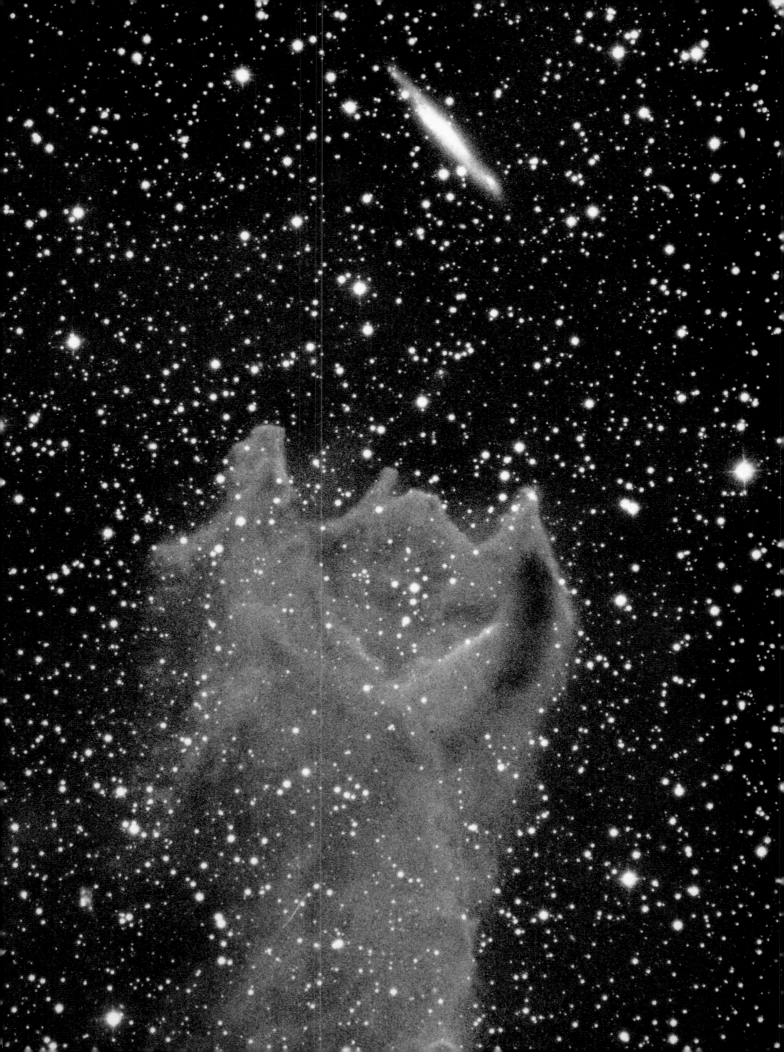

—— PLATE 26 ——

LIGHT ECHOES ROUND SUPERNOVA 1987A

From starbirth to stardeath. The elements from which the Earth and everything on it, including the material of your own body, are made were manufactured inside stars which exploded at the ends of their lives and scattered the material into space, where it formed the raw material for stars like the Sun and planets like the Earth. The most violent of these stellar explosions are known as Type II supernovas. They occur when a star more than ten times as massive as our Sun has exhausted all of its nuclear fuel, and can no longer produce the energy needed to hold itself up against the inward pull of gravity. It is as if the floor were pulled from under the star, and the whole core of the star suddenly collapses into a ball only a few kilometres across (perhaps even into a black hole). This releases a blast of gravitational energy so powerful that in a few seconds the star releases ten times as much energy as the Sun will release in 10 billion years of quiet nuclear burning. This blasts the outer layers of the star away into space, at speeds of up to 10,000 km per second.

The most recent nearby supernova of this kind was seen in the Large Magellanic Cloud in 1987, and is known as Supernova 1987A. The supernova occurred in the 30 Doradus complex, just below the Tarantula Nebula (see page 71). Seven years later, the HST took this picture of the region where the supernova occurred. The two large glowing rings of material lie in front of and behind the site of the supernova itself; it is thought that they are part of a spherical shell of gas blasted out in the explosion and expanding away from where it occurred. The two rings are being illuminated by radiation from a tiny neutron star that was a companion of the supernova before it exploded. Beams of radiation from near the poles of the spinning neutron star are sweeping round on either side like two lighthouse beams; although the beams are invisible, we can see their trace where they sweep across the gas from the supernova and energise it enough to make it glow.

The two white dots in the image are foreground stars in our own Galaxy; the inner bright ring marks a second shell of hot material expanding away more slowly from the site of the explosion.

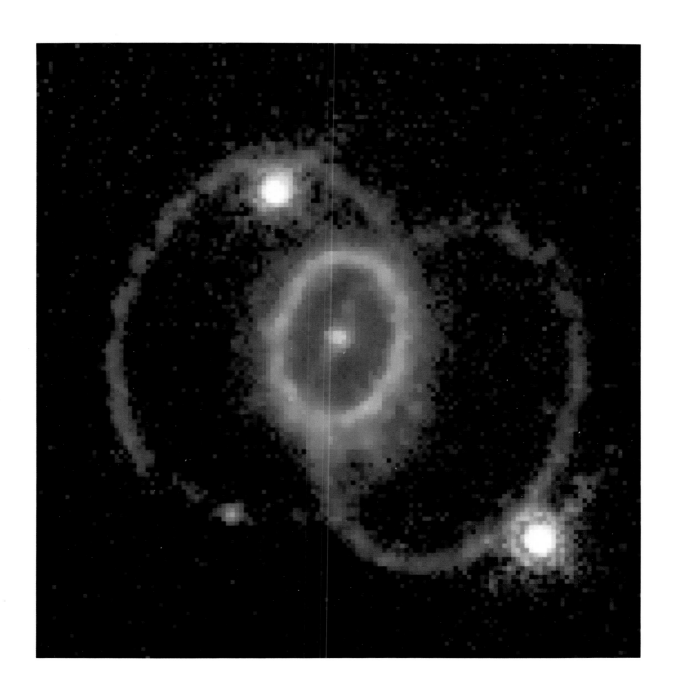

—— PLATE 27 ——

SUPERNOVA 1987A BEFORE AND AFTER

This pair of AAT photographs gives some indication of just how bright a supernova is. The picture at the top of the page is a routine photograph of part of the Large Magellanic Cloud, taken before the supernova seen in February 1987 occurred. The image below is a picture of the same part of the LMC taken soon after Supernova 1987A was discovered, but when the exploding star had already just started to fade from its brightest glory. The arrow in the top part of the picture marks the star that exploded – a seemingly ordinary large blue star catalogued as Sanduleak –69° 202.

At its brightest, for several days a supernova like this will shine as brightly as an entire galaxy of a hundred billion stars put together.

—— PLATE 28 ——

THE CRAB NEBULA

A supernova even closer to home than Supernova 1987A was seen in our own Galaxy in AD 1054, before the invention of the telescope. It was observed by Chinese astronomers with the naked eye, and was so bright that it was visible in daylight for 23 days. The expanding cloud of material from that explosion is known as the Crab Nebula, and the material in the cloud is still moving outwards at a speed of about 1,500 kilometres per second. The nebula lies 7,000 light years away, in the constellation Taurus.

Earlier in the twentieth century, astronomers were puzzled when they discovered from studying photographs of the Crab Nebula (inset) that it must have been expanding at the same steady speed observed today for nine centuries – they had expected that the cloud would slow down as it moved away into space from the site of the supernova explosion. But the discovery of a pulsar in the heart of the Crab Nebula in 1967 explained what was going on. The Crab Pulsar is the remnant of the star that exploded, a cosmic cinder in the form of a neutron star in which about the same amount of mass as there is in the Sun is packed into a ball just 10 km across. The ball is spinning once every 0.033 seconds (that is, thirty times every second), and as it spins it is flicking around a beam of radio noise, like a superfast celestial lighthouse.

The 'radio lighthouse' is what we detect on Earth, and what tells us the pulsar is there. But the radio noise actually comes from charged particles (mainly electrons) which are being whirled around in the grip of the magnetic field of the neutron star, and are them-selves being flung out at about the speed of light to collide with the gas of the nebula. Changing the analogy, it is like a super-powerful cosmic lawn sprinkler (a pair of similar sprinklers is responsible for the double rings associated with the remains of Supernova 1987A; see page 81). It is the energy from the pulsar that keeps pushing the material in the Crab Nebula outwards at a steady rate – the spinning neutron star stores so much energy (like a huge flywheel) that its measured slow-down, 300 billionths of a second each day, is exactly right to account for the entire energy output of the nebula, which is now 10 light years across.

When the HST turned its Wide Field Planetary Camera (WF/PC2) on to the central region of the Crab Nebula, it found even more direct evidence of the influence of the pulsar. In this image, the pulsar itself can be seen as the left-hand star in the pair near the centre of the picture. The pulsar is surrounded by a complex mixture of knots and wisps in the gas of the Nebula, and you can clearly see a pattern of ripples, like the ripples made by a stone dropped in a pond, around the pulsar. Images like this taken in a sequence over a span of several months show that these ripples are moving outwards from the pulsar at half the speed of light, pushed by the intense beam of particles and radiation from its 'lighthouse'.

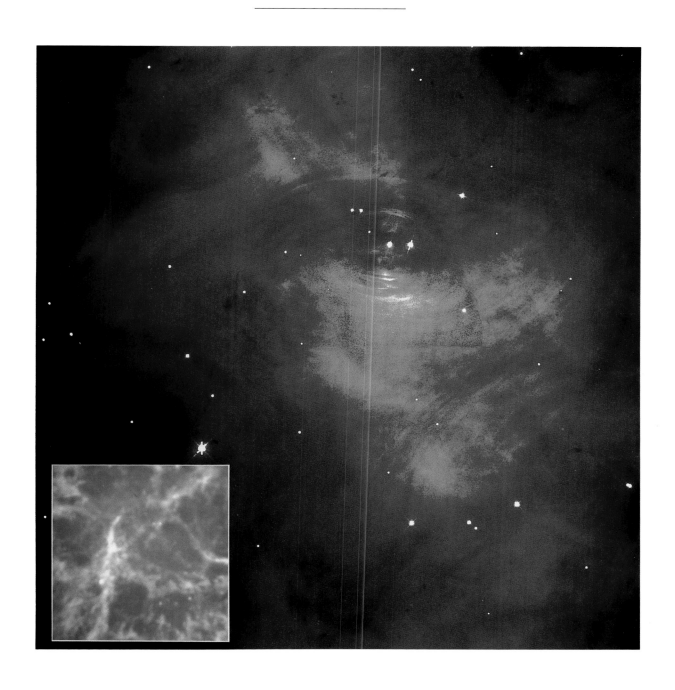

—— PLATE 29 ——

SUPERNOVA REMNANTS

This image shows two supernova remnants. Although they are superimposed on one another in this picture, this is simply because they happen to lie along the same line of sight. In fact, there is three times as much distance between the two remnants as there is between us and the nearer remnant. The smaller object, on the top right of the picture, is actually much further away long the line of sight. It is called the Puppis SNR, and is about 6,000 light years from Earth. Most of the image is filled with the much closer Vela SNR, which is only about 1,500 light years away, and has a diameter of about 230 light years. The light by which we see the Vela SNR today left it at about the time that the last vestiges of the Roman Empire in Italy were being overrun by invaders from the north. It took the remnant about 11,000 years to expand to this size from the supernova explosion in which it was born, so the supernova might have been seen by our ancestors around 9000 BC, about the time that agriculture was invented.

The image comes from X-ray observations made by the satellite ROSAT, launched in 1990. Although the two brightest parts of the Vela SNR can also be seen in optical light by ground-based telescopes, the X-ray observations were the first that showed the overall spherical structure of the remnant.

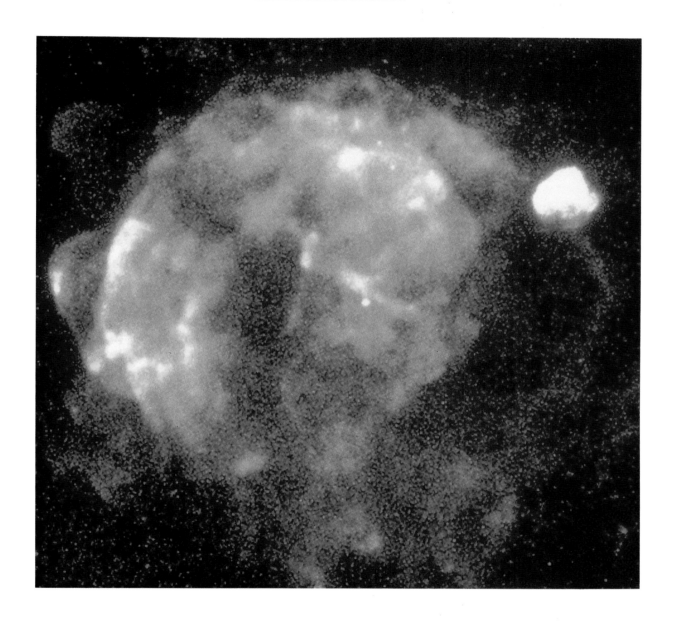

— PLATE 30 —

JUPITER AND THREE MOONS

The Sun and its family of planets formed together when a clump of gas and dust like the Bok globules in the image on page 73 collapsed in the wake of a supernova explosion which sent a blast wave rippling through the interstellar medium. Even though the cloud from which the Solar System formed was laced with heavier elements from previous stellar outbursts, the cloud still contained a mixture of 99.9 per cent hydrogen and helium (left over from the Big Bang) and only 0.1 per cent for everything else put together. In terms of the number of atoms around in the Solar System, sulphur is the tenth most common substance. For every atom of sulphur, there is 1 atom of iron, 2 atoms each of magnesium and neon, 3 atoms of silicon, 4 atoms of nitrogen, 20 atoms of carbon, 30 atoms of oxygen, 3,000 atoms of helium and 50,000 atoms of hydrogen. The Sun contains 99.86 per cent of the mass of the Solar System, and two-thirds of the remaining mass is concentrated in the giant planet Jupiter, pictured here in a photograph taken by the probe Voyager 1 in 1979.

Jupiter is five times further from the Sun than we are, and has 318 times as much mass as the Earth (0.1 per cent of the mass of the Sun). It is more than twice as big as all the other planets in the Solar System put together. The spacecraft was 28.4 million km from Jupiter when it obtained this image, which shows the Great Red Spot and three of Jupiter's moons. Io is silhouetted as a brown-yellow disk against the bulk of Jupiter, while Callisto looks like a blob in the northern hemisphere of the planet. Europa is below Jupiter in this image.

Counting the planets inwards from outside the Solar System, after passing through the realm of the comets a visitor from another star would encounter first Pluto, a small icy world far from the Sun, then the four gas giants Neptune, Uranus, Saturn and Jupiter, then a belt of rocky debris (the asteroid belt) and finally the four small, rocky planets Mars, Earth, Venus and, closest of all to the Sun, Mercury.

—— PLATE 31 ——

I O

Jupiter's moon Io is pictured here in an image obtained by the Galileo space probe in June 1996. Io is the innermost of the four large moons of Jupiter discovered by the astronomer Galileo early in the seventeenth century. It orbits only 422,000 km from the planet, and is locked in a gravitational embrace so that it always keeps the same face towards Jupiter (just as our Moon always keeps the same face towards the Earth). The diameter of Io is 3,630 km, and it has a mass one-fifth greater than that of our Moon. The strong tidal forces produced by Jupiter repeatedly squeeze and stretch the moon, heating its interior and making it the most volcanically active object in the Solar System. The yellowish-brown colour of the surface is a result of this activity, which spews sulphurous material out from within the moon; some of these sulphurous equivalents of lava flows stretch for 200 km from the volcanoes which gave them birth.

—— PLATE 32 ——

SATURN

Although it is smaller than Jupiter (its mass is only ninety-five times that of the Earth) and further out from the Sun (roughly twice as far), Saturn is one of the most intriguing and beautiful objects in the Solar System because of its prominent system of rings. This view of Saturn as a crescent, with its shadow eclipsing part of the ring system, was obtained from Voyager 1 in November 1980, four days after it had passed Saturn on its way out of the Solar System, looking back towards the planet. Just as the planet eclipses part of the rings, so the rings cast a shadow on part of the planet. The rings are not solid, but are made up of countless pieces of ice and rock, each orbiting the planet in its own trajectory. As well as the rings, Saturn has more than twenty moons. Like Jupiter, which has at least sixteen moons, in some ways it resembles a 'solar system' in miniature. It is likely that the planets themselves formed from a similar but much larger ring system round the young Sun, with the pieces of material in the rings colliding and sticking together, building up to make the planets.

The outermost of the gas giants, Uranus and Neptune, resemble Saturn, but without the prominent rings. As the name implies, they are largely composed of gas, chiefly hydrogen, much of which is locked up in compounds such as methane and ammonia.

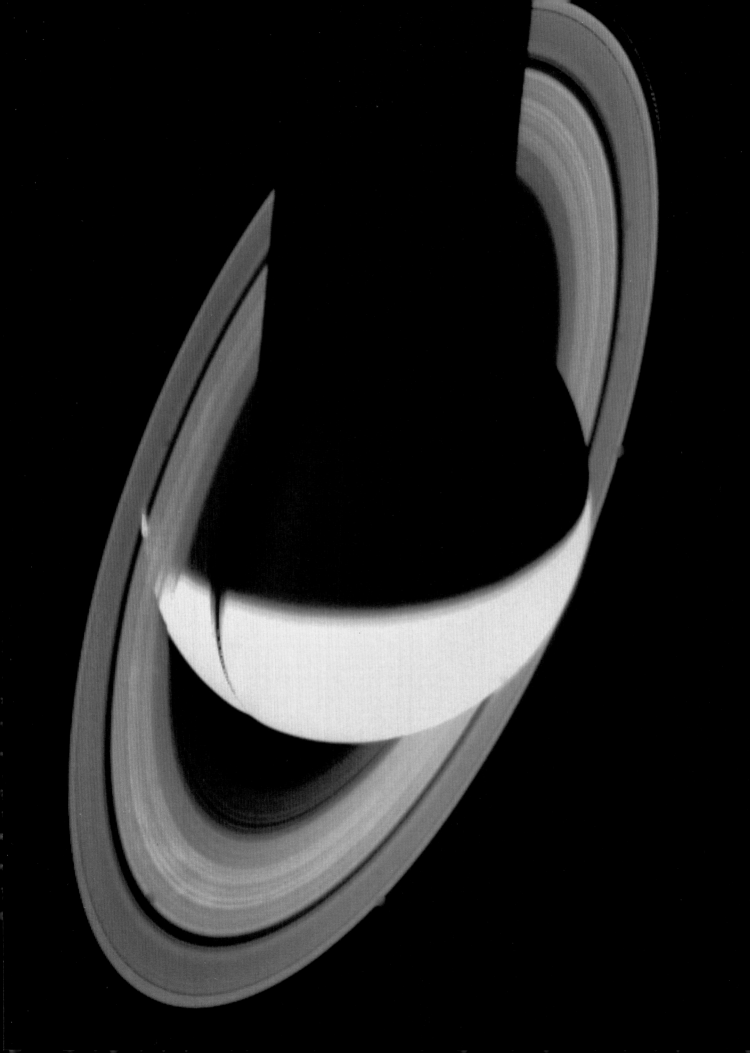

——— PLATE 33 ———

MARS

The first of the small, rocky planets that our visitor from another star would encounter on a journey towards the centre of the Solar System is Mars, the red planet. It isn't really quite as red as this, but the colour has been enhanced in this picture, obtained by the Viking Orbiter mission, to bring out the detail.

Mars has a mass only a little more than one-tenth of that of the Earth, and a diameter roughly half that of the Earth. It orbits the Sun once every 686.98 of our days, at an average distance of one and a half times further from the Sun than we are. It has a thin atmosphere of carbon dioxide, and turns on its axis once every 24 hours, 37 minutes and 23 seconds, giving it a day about the same length as ours. This is a coincidence.

In terms of origins, the most important thing to notice about Mars is the battered state of its surface, smothered in impact craters. Because the atmosphere of Mars is thin, there has been much less weathering of the surface than occurs on Earth. And because there is no life on Mars, the surface features have not been obliterated by vegetation. What we see here is evidence, in the form of the record in the rocks, of the last great phase of planet building, when rocks rained down on Mars about 4 billion years ago, as it swept up the last of the debris in its orbit.

—— PLATE 34 ——

VENUS

The surface of Venus is very different from that of Mars. From a comparison between the number of craters on Venus and the number of craters seen on the surfaces of the Moon (page 109) and Mercury (page 99), astronomers infer that the whole surface of Venus was turned over some 600 million years ago, in a great cataclysm in which so much lava flowed out from its interior through cracks in its crust that the entire planet was resurfaced with new rock. All the craters we see today on the surface of Venus were formed by impacts within the past 600 million years; this discovery has some bearing on the understanding of our own origins (see page 111).

Venus has a very dense carbon dioxide atmosphere, and is completely shrouded with clouds rich in sulphuric acid, so no information about its surface was available until astronomers developed techniques to map the planet by radar. This image is mainly a result of radar mapping by the space probe Magellan, which went into orbit around Venus and charted 98 per cent of its surface. The gaps have been filled in using data from earlier missions, both Soviet and American, with some data coming from Earth-based radar observations using the Arecibo radio telescope in Puerto Rico. The colours used to produce this image are based on the actual colours at the surface of Venus seen by the Soviet probes Venera 13 and Venera 14, which landed on the planet and sent back data briefly before being simultaneously cooked, corroded and crushed by the enormous pressure at the surface – ninety times the atmospheric pressure at sea level on Earth.

The planet's highest mountain range, Maxwell Montes, is the bright feature just below the middle of this image; it rises 11 km above its surroundings. The image also shows lava flows, impact craters, ridges and other structures.

Venus is the second planet out from the Sun. It has 82 per cent as much mass as the Earth, and orbits the Sun once every 225 days at, on average, 72 per cent of our distance from the Sun. It rotates very slowly, once every 243 of our days. Because of a combination of its proximity to the Sun and the strong greenhouse effect of its thick carbon dioxide atmosphere, the surface temperature of Venus is above 450 degrees C. The surface is a searing, lifeless desert, scoured by strong winds and showers of acid rain.

PLATE 35

MERCURY

Mercury is the innermost planet of the Solar System, orbiting the Sun at only 39 per cent of the distance from the Earth to the Sun, and taking only 87.97 of our days to complete each orbit. It turns on its own axis once every 58.64 of our days, so that three 'days' on Mercury last for two of the planet's 'years'. Its mass is only 5 per cent of that of the Earth, and its diameter is 4,880 km, making it intermediate in size between the Moon and Mars. It has no atmosphere at all, so there has never been any weathering of its surface. And, unlike Venus, its surface has never been overturned in a cataclysmic upheaval. So the surface of Mercury carries a reminder of every blow from space that has ever struck it, and provides the best record we have of just how intense the bombardment of the inner planets has been. This record is clearly shown in this image, a mosaic of pictures sent back by the probe Mariner 10, which flew past Mercury in 1974. The image has been highlighted in false colour to show the details.

The most spectacular impact feature on Mercury is the Caloris Basin, about 1,300 km across and surrounded by a ring of mountains 2 km high, created in the impact. The basin is big enough to contain the British Isles. The evidence of extensive cratering right across the inner part of the Solar System, from Mars to Mercury, shows that the bombardment which occurred 4 billion years ago was not restricted to just one planet, but happened right across this region as the rocky planets formed by sweeping up cosmic debris. The cratering of Venus over the past 600 million years, though, shows that there are still traces of that debris around in our part of the Solar System, and that from time to time such rocks from space still strike the inner planets – including the Earth (see page 113).

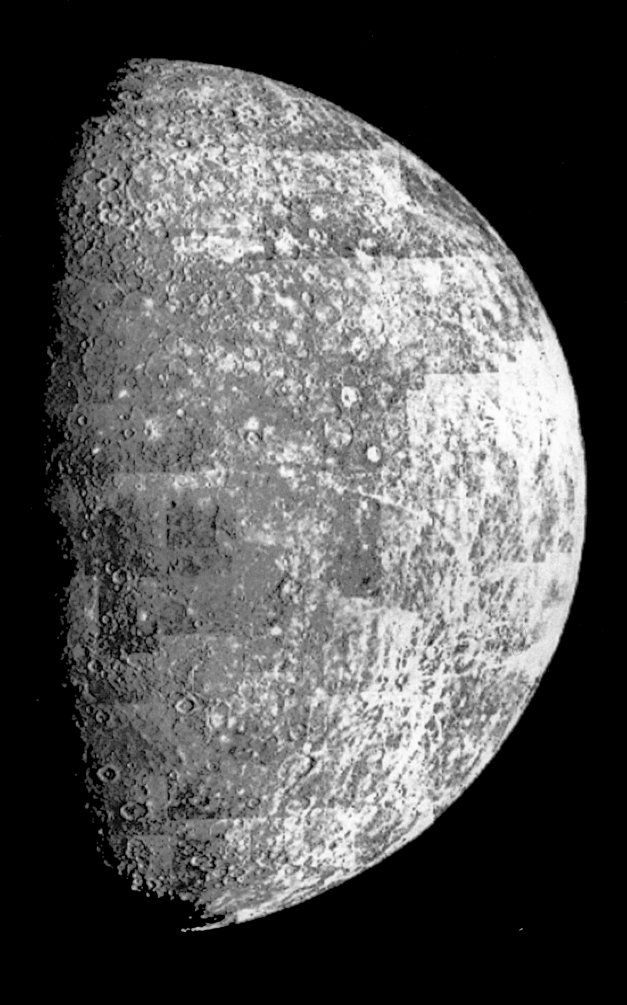

—— PLATE 36 ——

THE SUN

Our Sun, which sits at the centre of the Solar System, is an ordinary star, roughly halfway through its lifetime on the main sequence. It only looks big and bright to us, compared with other stars, because it is so close – the Sun is 150 million km away, but the next nearest star is more than 4 light years away. The Sun is a ball of hot gas (roughly three-quarters hydrogen and one-quarter helium) with 330,000 times as much mass as the Earth and a diameter 109 times the diameter of the Earth, corresponding to a volume rather more than a million times the volume of the Earth. This gives it an overall density of a third of the Earth's density, only 1.4 times the density of water. But that average disguises the fact that the outer layers of the Sun are very tenuous, while the innermost core is very dense.

At the heart of the Sun, its material is packed together so tightly that it has 160 times the density of water (twelve times the density of lead). But instead of being made of atoms, this core material is made of nuclei of hydrogen and helium, atomic kernels stripped of their electrons, and even at these densities it behaves like a gas. The density drops off slightly over the inner 1.5 per cent of the Sun (15,000 times the volume of the Earth), which actually contains half of its total mass (roughly 150,000 times the mass of the Earth). The temperature at the centre of this core is about 15 million degrees C, and the pressure is 300 billion times the atmospheric pressure at the surface of the Earth.

Above this core, the density drops rapidly, to be the same as water halfway to the surface of the Sun, and as thin as the air we breathe two-thirds of the way to the surface. In the top 10 per cent of the Sun, the density is less than 1 per cent of the density of water. At the visible surface of the Sun, the temperature is only 6,000 degrees C. But as this image shows, the Sun extends into space beyond the visible surface, through a tenuous layer known as the chromosphere into the corona, which extends for millions of kilometres into space, and blends into the so-called solar wind of particles that stream outwards from the Sun and (among other things) cause the auroras here on Earth.

This image is of the Sun as you will never see it, observed in 1996 by detectors on board the SOHO satellite, sensitive to extreme ultraviolet radiation. The green colour has been chosen arbitrarily to bring out the detail, which includes flaring activity on the surface of the Sun and a very clear system of arching loops, called prominences, on the left-hand side of the picture. The Earth would fit comfortably under one of these arches, which are temporary features linked to the activity of the Sun's magnetic field.

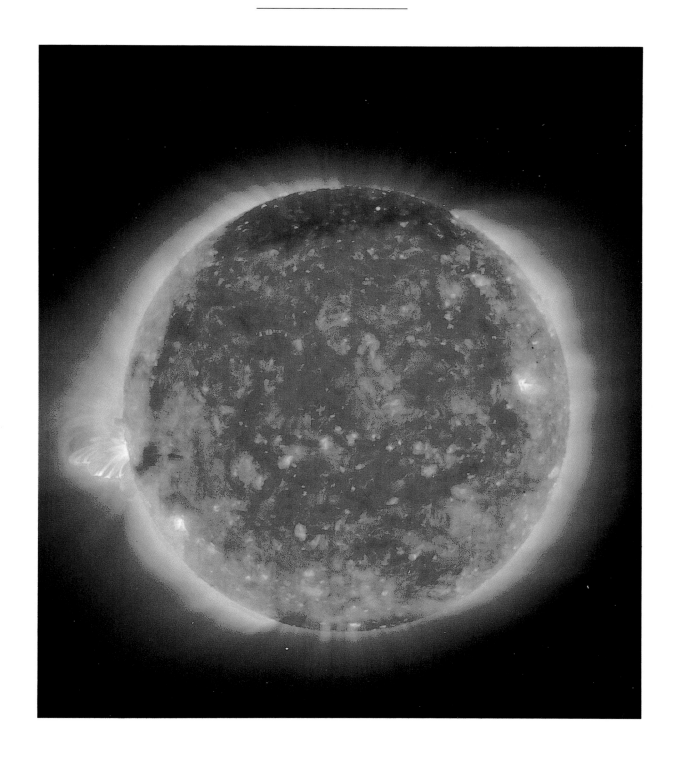

—— PLATE 37 ——

HALLEY'S COMET

The image most of us have of a comet is like the photogenic visitor from space pictured on page 23. But what causes this heavenly display? In 1986, the space probe Giotto carried out a close encounter with Halley's Comet, and sent back this picture of the lump of material at its head. A comet is, in fact, a lump of icy material and dust (a 'dirty snow-ball') which spends most of its life in the outer reaches of the Solar System, where it is undetectable. It is only when a comet is disturbed into an orbit that brings it into the inner Solar System, where it is heated by the rays of the Sun, that material begins to boil off from its surface (as you can see happening at the top left of this picture) and create the spectacular tail that gives the characteristic appearance of a comet seen from Earth.

There are thought to be billions of these cometary nuclei, as they are called, in a cloud orbiting the Sun in leisurely fashion literally halfway to the nearest star, far beyond the orbit of Pluto. They are bits of primordial material left over from the origin of the Solar System. The mass of all these comets put together adds up to about a hundred times the mass of the Earth. When, occasionally, one or more of these objects gets disturbed (perhaps by the gravity of a passing star), and falls into the inner Solar System, some (like Halley's Comet) are captured by the gravity of Jupiter into an orbit which brings them looping past the Sun repeatedly. Halley's Comet itself has an orbit 76 years long, taking it out beyond the orbit of Neptune, and in to within the orbit of Venus.

The nucleus of Halley's Comet is shaped like a lumpy potato, about 16 km long and 8 km wide, similar to the size of the island of Manhattan. About 80 per cent of its mass is in the form of water ice, and the rest is made up of lumps of rock embedded in the ice. The whole thing is covered in a black layer of carbon dust (soot), and has a mass of 300 billion tonnes. Comets in orbits like that of Halley's Comet cross the orbit of the Earth, so if the Earth is in the right (or wrong!) part of its orbit at the right time, it can get struck by such lumps of cosmic debris. It is almost certain that the catastrophe that killed the dinosaurs and many other forms of life on Earth 65 million years ago was due to an impact of this kind.

—— PLATE 38 ——

ASTEROID IDA AND ITS MOON

Comets are not the only kinds of debris still floating around the Solar System, left over from the material out of which the planets were made. Very many lumps of rocky material, parts of a proto-planet that failed to form, orbit in a band known as the asteroid belt (like the rings of Saturn on a very large scale) between the orbits of Jupiter and Mars. The reason why they failed to form a planet is the disturbing influence of Jupiter's gravitational pull. More than 5,000 members of this asteroid belt have been identified and catalogued. There are probably half a million of them big enough to be seen using the 200-inch (5-metre) telescope at Mount Palomar (if anyone could be bothered to look for them), and many millions more with sizes of a few hundred metres.

This image shows a typical member of the asteroid belt, known as 243 Ida, photographed by the space probe Galileo in August 1993 on its way to Jupiter (the number means that Ida was the 243rd asteroid to be identified). This asteroid is about 58 km long and 21 km wide. The Galileo images also revealed, for the first time, that Ida has a tiny moon, dubbed Dactyl, visible at the top of this picture. Dactyl is only 1.6 km long and 1.2 km wide, and orbits Ida at a distance of 90 km. By measuring the orbit of Dactyl around Ida, astronomers can work out the mass of the asteroid; Ida has a mass of about 70 thousand billion tonnes.

The cameras on board Galileo were sensitive to infrared light, and to human eyes the asteroid would appear a dull grey. The colours in this image show details in the surface that would be missed by unaided human eyes, such as the bright bluish areas around some of the craters, which show a different amount of iron in these regions.

The orbits of the asteroids are still being perturbed by the gravity of Jupiter, so from time to time a member of the asteroid belt is disturbed into an orbit which takes it closer to the Sun. Some of these asteroids go so close to the Sun that they cross the orbit of the Earth, and they will stay in such an orbit until they hit something. Several such 'Earth-crossing' asteroids have been discovered, and there are certainly more that are too faint to be seen.

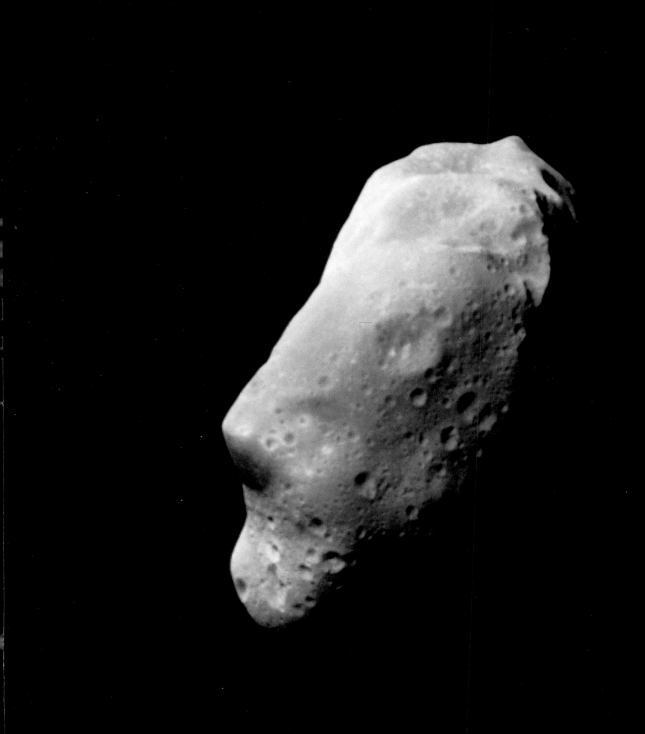

—— PLATE 39 ——

THE EARTH AND THE MOON

We have saved the best for last. This is our home in space, the Earth and the Moon photographed by the Galileo space probe in December 1992, looking back to home on its way to Jupiter. The picture was taken from a distance of 6.2 million km; the bright Earth reflects three times as much sunlight as the Moon does. Antarctica is just visible through the clouds on Earth, but the Moon has an unfamiliar appearance because we are looking at the side that is always turned away from the Earth. The shadowy indentation that can just be discerned in the surface of the Moon near to the boundary between its dark and sunlit sides (the line of dawn on the Moon) is the Aitken Basin, at the lunar South Pole. This is one of the largest and oldest impact features on the Moon. It is 13 km deep and 2,500 km across, and was created by the glancing impact of an asteroid 200 km across.

—— PLATE 40 ——

THE MOON

The battered face of the Moon is shown in detail in this false-colour image obtained by the Galileo space probe in December 1992. The part of the Moon visible from Earth is on the left-hand side of this picture. Bright, pinkish areas are the highland material, such as the mountains around the oval Crisium impact basin, a lava-filled crater at the bottom left of the picture. The solidified lava flows are coloured in shades from blue to orange. To the left of Crisium, the Mare Tranquillitatis shows up dark blue, because the lava there is particularly rich in titanium. Thin layers of material spread by relatively recent impacts show up light blue; the youngest craters have bright blue rays extending out from them.

Because the Moon is airless, there has been no weathering; and because it is not geologically active, the surface has not been overturned like the surface of Venus (see page 97). What we see on the surface of the Moon today is a record of the battering it has received over the past 4 billion years or so. We know that this was not a unique battering caused by the way the Moon formed in orbit around the Earth, because similar craters are seen on the surfaces of Mercury, Venus and Mars. So there is no doubt that the Earth has received a similar pounding down the aeons.

Most of the battering occurred about 4 billion years ago, in the last stages of the formation of the Moon and the inner planets of the Solar System. The Tranquillitatis lava flow, for example, is 3.8 billion years old, and it is among the youngest material on the surface of the Moon. But there are many younger craters scarring the surface of such ancient lunar lava flows. If you were to draw a circle 100 km in diameter at random anywhere on the surface of the Moon you would find 500 craters at least a kilometre in diameter, all younger than 600 million years. From this evidence, and from counts of the numbers of craters on Mercury, Venus and Mars, astronomers calculate that an object bigger than 1 km across hits the Earth (which is a much bigger target than the Moon) once every 200,000 years or so, releasing the energy equivalent of the explosion of a 20,000 megatonne nuclear bomb (400 times more powerful than any bomb ever exploded on Earth), and producing a crater 20 km across. An object big enough to make a crater 100 km across strikes our planet roughly every 50 million years.

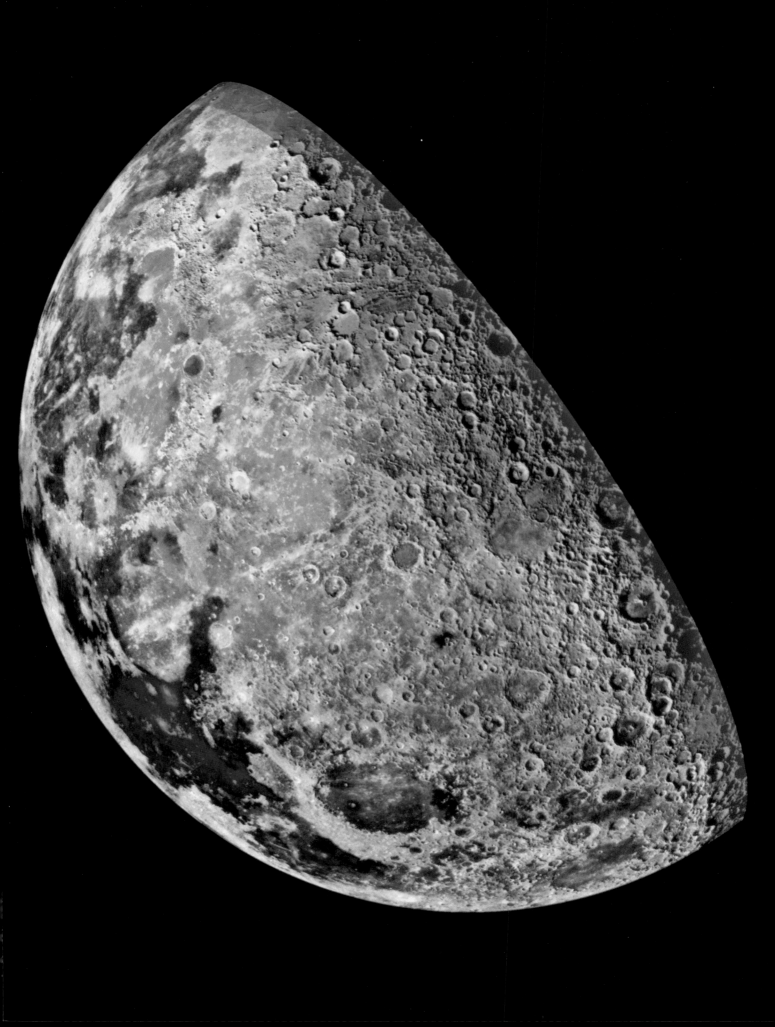

—— PLATE 41 ——

THE EARTH

It is hard to believe that the tranquil Earth, floating peacefully in space, is subject to a rain of cosmic missiles, and has been battered by impacts as much as the surface of the Moon or any of the other inner planets of the Solar System. The evidence is not obvious because even over the third of the Earth that is not covered by water, wind and weather have worn away the craters from the impacts; and although our planet has not been subjected to the same kind of cataclysmic upheaval that occurred on Venus 600 million years ago, the surface layer of the Earth is constantly being recycled, spreading out from ocean ridges and being pushed back down into the interior of the planet in deep trenches along the edges of some continents. As a result, none of the present-day sea floor is more than 200 million years old, and even the continents are not as old as Mare Tranquillitatis (see page 109). In addition, most of the impact sites that are still present on the surface of the Earth today are covered by vegetation, and do not show up as crisply as equivalent features on the surface of the Moon.

This image of the Earth was obtained by our old friend the Galileo space probe, looking back from a distance of about 1.3 million miles. South America is in the centre of the picture, with Antarctica clearly visible to the south, and weather fronts trail across the South Atlantic Ocean.

—— PLATE 42 ——

COSMIC IMPACT ON EARTH

Observations from space show that the surface of the Earth really is scarred with the traces of many impacts by objects from space, known collectively as meteorites. Both photographs and radar images from space highlight features that are not obvious from the ground, or even from aircraft. This particular impact crater, in southwest Namibia, is shown in an image obtained from a space radar system carried on a flight of the space shuttle *Endeavour*, on 14 April 1994. The radar data have been converted into colours which highlight the structure of the crater, which is a relatively modest example of its kind, only 2.5 km wide and 130 metres deep (but the true bottom of the crater lies a further 100 metres below the surface of the sand that has partly filled it). It was formed by a meteorite that hit the Earth about 5 million years ago, just before our ancestors learned to walk upright. This crater is apparent at the surface of the Earth, but only the radar images reveal the detail of the structure and its full extent.

An impact like this in a populated region of the globe today would be a major disaster. The impact that wiped out the dinosaurs, 65 million years ago, produced a crater 180 km in diameter, which was buried by geological activity and has only recently been identified, in what is now the Yucatan Peninsula of Mexico. This is one of the biggest craters yet identified on the surface of the Earth (an impact that big occurs only every 100 million years or so), and the event that formed it was a terminal disaster as far as the dinosaurs and many other species were concerned.

But the demise of the dinosaurs opened the way for some creatures that survived the event to recover and thrive in a world free from dinosaurs. Among the most successful of these survivors were the mammals, our ancestors, to whom (with hindsight) that great meteorite impact can be seen as a lucky break. We owe our own origins directly to the blow from space that removed the dinosaurs from their dominant position on Earth, and opened up opportunities for other forms of life. But how those survivors from 65 million years ago survived and evolved to become us is another story.

GLOSSARY

Anglo-Australian Telescope A 4-metre telescope opened in 1971. Based at Siding Spring in eastern Australia, the AAT is able to observe a wealth of interesting objects unobservable from the northern hemisphere, such as the Galactic centre and the Large and Small Magellanic Clouds.

Astronomical unit (AU) The average distance between the Earth and the Sun, about 150 million km or 93 million miles. The AU is a useful unit for describing the distances between planets.

Big Bang The hypothetical beginning of the Universe in which the whole of space and time was created in an infinitesimally small point which has expanded outwards for the past 10–20 billion years into the Universe we know today. The phrase 'Big Bang' was originally coined by astronomer Fred Hoyle as a term of derision.

Black hole An object that has been shrunk by gravity until its gravitational pull is so strong that nothing, not even light, can travel fast enough to escape from it.

Cepheid variable A type of very bright star which pulses with a regular period. If the period of the pulses is known then the distance to the star can be calculated very accurately. Hence Cepheids are very useful in finding the distances to galaxies.

COBE The COsmic Background Explorer satellite, launched in 1989 to map the fluctuations in the radiation left over after the Big Bang. As the Universe has expanded, this radiation has cooled to a temperature of only 3 degrees K (–270 degrees C).

Comet A small body in the Solar System made up of ice and rock. Comets become visible as they near the Sun because of their tail, which is made up of material boiled off the comet by the heat of the Sun.

Dark matter The name given to any matter that cannot be seen. As we cannot see it we do not know what it is, but we do know that it is there. As much as 90 per cent of the Universe may be made up of mysterious dark matter.

Disk galaxy see **spiral galaxy.**

Elliptical galaxy A round, featureless galaxy which, unlike a spiral galaxy or irregular galaxy, does not contain gas and so cannot form new stars.

ESA The European Space Agency, which builds and launches satellites and planetary missions. In the modern spirit of international cooperation (mainly because space missions are becoming too expensive for one country alone) ESA collaborates extensively with NASA on projects such as SOHO, Giotto and the Hubble Space Telescope.

Galaxy A collection of maybe hundreds of billions of stars, and often gas, held together by gravity.

Galaxy cluster A group of galaxies bound together by the force of gravity. Galaxy clusters may contain thousands of individual galaxies.

Galileo The Galileo spacecraft (named, unsurprisingly, after Galileo Galilei), launched from the space shuttle *Atlantis* in October 1989. Following an unusual orbit, Galileo first travelled towards the Sun and used the gravity of Venus and Earth as a boost in order to reach Jupiter in late 1995.

Giotto A joint NASA/ESA mission to study two comets, most importantly Halley's Comet. Launched in July 1985, Giotto passed within 600 km of the nucleus of Halley's Comet on 13 March 1986. In July 1992 Giotto also passed within 200 km of the comet Grigg-Skjellerup.

Gravitational lens When light from a distant galaxy is bent by the gravitational pull of another galaxy or galaxy cluster we can sometimes see multiple images of the distant galaxy. This effect is gravitational lensing.

Infrared The part of the spectrum with frequencies just too long for the human eye to see.

Irregular galaxy A galaxy which has no clear structure. Irregular galaxies seem to have been much more common early in the history of the Universe.

Light year The distance travelled by light in one year. Light travels at 300,000 km per second, and in one year covers 9.46 thousand billion km or 5.88 thousand billion miles.

Mariner 10 A NASA spacecraft which explored the inner Solar System in the mid-1970s. Mariner 10 has provided the most detailed information we have on Mercury, the innermost planet, in a series of three fly-bys, the closest passing only 327 km above the surface.

NASA The National Aeronautics and Space Administration, the body that runs the USA's space programme. It is without doubt the best-known and most successful space programme in the world.

Nebula A huge cloud of dust and gas, often weighing many millions of times more than the Sun. Frequently found to be the birthplace of new stars.

Neutron star Very massive stars (greater than about eight times the mass of the Sun) end their lives as a supernova when the outer layers of the star are blown away in a massive explosion. What is left is usually a neutron star – a densely packed ball of neutrons a few kilometres across with a mass greater than that of the Sun.

NGC numbers The NGC number is the classification number from the New General Catalogue, first published in 1888. Many objects in the sky are known by their NGC number.

Planetary nebula At the end of a star's red giant phase its outer layers are blown away, creating a shell of dust and gas around the star known as a planetary nebula. (Planetary nebulas have nothing to do with planets, although they were once thought to, hence their name.)

Pulsar A pulsar is a very rapidly spinning neutron star (possibly revolving thousands of times per second) with an incredibly strong magnetic field. Every time the pulsar spins round this magnetic field beams a strong radio pulse with such regularity that when pulsars were discovered they were thought to be evidence of alien intelligence.

Quasar The core of a distant galaxy which gives out incredible amounts of energy. This energy is thought to be produced around a huge black hole in the centre of this type of galaxy, which is billions of times heavier than the Sun.

Red giant When a star has burnt all of the hydrogen in its core, it expands before starting to burn helium. This expansion cools the surface of the star and it becomes a red giant.

ROSAT The ROentgen SATellite, an X-ray satellite named after the discoverer of X-rays. Launched in June 1990, one of the main objectives of the ROSAT mission was to map the entire sky in X-rays.

SOHO The SOlar and Heliospheric Observatory, designed to observe the Sun continuously in ultraviolet light, looking for activity such as solar flares and sunspots.

Spectrum When light is split up into its component colours, the result is a spectrum. In the spectra of astronomical objects, some thin bands of colour are found to be missing. These missing lines provide a great deal of information on the composition of an astronomical object which could not be found in any other way.

Spiral galaxy A galaxy like the Milky Way, which has a thin disk of stars thousands of light years in diameter. In this disk new stars are forming in spiral patterns, which shine brightly and give this type of galaxy their name.

Star A mass of mostly hydrogen and helium gas which collapses together under gravity. As a star collapses it gets hot enough in the centre to begin the nuclear fusion of hydrogen into helium. This reaction creates huge amounts of energy which keeps the star shining.

Supernova When a star more than ten times as massive as the Sun has burnt all of the hydrogen in its centre it does not expand to become a red giant, but violently ejects its outer layers in a supernova explosion, which can be seen from millions of light years away.

Ultraviolet The part of the spectrum with frequencies just too small for the human eye to see. The atmosphere blocks out the vast majority of the ultraviolet radiation from space, especially that from the Sun which can be extremely harmful.

Viking mission The Viking 1 and 2 spacecraft were launched in 1975 to explore Mars. Each Viking contained a lander which parachuted to the surface of Mars and an orbiter which took high-resolution images of the planet's surface from orbit. The Vikings found considerable evidence that water was once common on the surface of Mars, but failed to find any evidence for life there now.

Voyager mission The last of NASA's Mariner series of planetary explorers; Voyagers 1 and 2 were launched in late 1977 (Voyager 2 actually being launched first although it arrived at Jupiter and Saturn later). The Voyager probes were designed to explore the outer Solar System, in particular the giant planets. The Voyager missions have only just been shut down.

White dwarf The final fate of most stars is to become a white dwarf, a very dense, hot star about the size of the Earth, which slowly cools down until it is dark.

Wide Field and Planetary Camera (WF/PC) An instrument carried on the Hubble Space Telescope, consisting of four cameras which can image a wide area in the sky in great detail. The original WF/PC1 was replaced during the 1993 HST repair mission with the improved WF/PC2, containing extra optics to counter the flaw in the HST's mirror.

X-ray A very energetic form of electromagnetic radiation which can be very harmful. Best known for their medical uses, X-rays are produced in the hot outer layers of the Sun and are (thankfully) blocked from reaching us by the atmosphere.

PICTURE CREDITS

INTRODUCTION

Star trails. CREDIT: Copyright Anglo-Australian Observatory. Photograph by David Malin.

The 100-inch Hooker Telescope. CREDIT: The Observatories of the Carnegie Institution of Washington.

Edwin Hubble and the Hooker Telescope. CREDIT: The observatories of the Carnegie Institution of Washington.

The Hubble Space Telescope. CREDIT: NASA.

Distant irregular galaxies. CREDIT: Richard Griffiths (JHU), The Medium Deep Survey Team and NASA.

COBE all-sky map. CREDIT: NASA Goddard Space Flight Center and the COBE Science Working Group.

The COBE satellite. CREDIT: NASA Goddard Space Flight Center and the COBE Science Working Group.

The spiral galaxy M100. CREDIT: NASA.

Comet Hyakutake. CREDIT: Copyright Luis Chinarro.

The Orion Nebula. CREDIT: C. R. O'Dell (Rice University) and NASA.

The Horsehead Nebula. CREDIT: Copyright Anglo-Australian Observatory. Photograph by David Malin.

PLATES

1. The COBE four-year map. The total data taken over the COBE mission on the fluctuations in the microwave background was assembled to produce the COBE four-year map of the sky. This is the best information we have on the structure of this radiation across the whole sky, although a new mission is planned to improve on our knowledge. CREDIT: NASA Goddard Space Flight Center and the COBE Science Working Group.
2. Galaxy formation simulation. This picture shows the results of a huge supercomputer simulation carried out by the Virgo Consortium, an international group of astronomers that studies the evolution of structure in the Universe. The region of the Universe simulated in the computer world is around 2,000 trillion cubic light years in extent! CREDIT: The Virgo Consortium.
3. The Lick galaxy map. This map shows the brightest million or so galaxies visible from the northern hemisphere. The Universe is estimated to contain over a billion galaxies visible to the most powerful telescopes. Many of these galaxies are grouped into clusters, superclusters and walls of galaxies with huge voids between them. CREDIT: M. Seldner, B.L. Siebers, E.J. Groth and P.J.E. Peebles (*Astronomical Journal*, **82**, 249, 1977).
4. The Coma galaxy cluster. A picture taken in X-rays by ROSAT of one of the closest giant clusters of galaxies. The hot intergalactic gas that appears in this picture has a total mass probably far greater than that of all the individual galaxies put together. In addition there may be exotic invisible matter in

the cluster which has even more mass than the gas. CREDIT: S.L. Snowden, Universities Space Research Association and NASA Goddard Space Flight Center.

5. The homes of quasars. All of these images were taken with the Hubble Space Telescope's WF/PC2. They reveal the faint host galaxies of these quasars which would be invisible to telescopes on Earth, allowing a study of the families of host galaxies impossible before. CREDIT: John Bahcall (Institute for Advanced Study, Princeton), Mike Disney (University of Wales) and NASA.

6. The Hubble Deep Field. This picture is formed from many WF/PC2 images taken in four colours (three in the visible spectrum and one in infrared) over 150 orbits of the Hubble Space Telescope. CREDIT: Robert Williams and the Hubble Deep Field Team (STScI) and NASA.

7. A very young galaxy in the Hubble Deep Field. This small portion of the Hubble Deep Field shows a small and unremarkable red blob at its centre. This is a distant galaxy that appears only in the infrared images. This is because the galaxy is so distant that its light has been redshifted by the expansion of the Universe so that it is unseen in visible light (the infrared light we now see was once ultraviolet light). CREDIT: Ken Lanzetta and Amos Yahil (State University of New York at Stony Brook) and NASA.

8. A gravitational lens in the galaxy cluster Cl0024+1654. By combining WF/PC2 images in red and blue light taken on 14 October 1994 this picture of the gravitational lensing of light through a galaxy cluster was created. CREDIT: W.N. Colley and E. Turner (Princeton University), J.A. Tyson (Bell Labs, Lucent Technologies) and NASA.

9. The massive elliptical galaxy M87. In this picture it appears peaceful. M87, however, is secretly active, producing huge radio jets from its centre and generating thirty times more energy output in X-rays than in visible light. It has also probably attained its huge size by cannibalising many other galaxies. CREDIT: Copyright Anglo-Australian Observatory. Photograph by David Malin.

10. Face-on spiral NGC 2997. If we were able to go outside the Milky Way and look at it from above, it would probably look very much like NGC 2997. Spiral galaxies such as the Milky Way, M101 and NGC 2997 settled into their disk shapes early in their histories. The fact that they are still disks tells us that they have not had any violent encounters with other large galaxies as that would have destroyed their fragile structures. CREDIT: Copyright Anglo-Australian Observatory. Photograph by David Malin.

11. Edge-on spiral NGC 4945. Again, NGC 4945 looks very much as our own Milky Way would look if viewed side-on. The Sun would be found very near the plane of the disk, about two-thirds of the way out from the centre. CREDIT: Copyright Anglo-Australian Observatory. Photograph by David Malin.

12. Barred spiral NGC 1365. Although barred and unbarred spiral galaxies can look quite different, they are basically the same sort of galaxy. Indeed, there are theories that say that bars tend to appear in spiral galaxies without them, but as soon as they do appear they become unstable and disappear again. If this is so, whether a spiral is barred or unbarred just depends on when you look at it. CREDIT: Copyright Anglo-Australian Observatory. Photograph by David Malin.

13. Starburst irregular galaxy NGC 1313. This galaxy lies fairly close to us, at a distance of only about 15 million light years, and was the site of Supernova 1978k (the eleventh supernova in 1978), which was discovered from its X-ray emission rather than from its visible light. CREDIT: Copyright Anglo-Australian Observatory. Photography by David Malin.

14. The spiral galaxy M101 in ultraviolet light. The space shuttle *Endeavour* carried the Ultraviolet Imaging Telescope during the Astro-2 mission, which was used to image a number of interesting astronomical objects. M101 is one of the largest spiral galaxies – its diameter is nearly three times that of the Milky Way. CREDIT: NASA.

15. The Large Magellanic Cloud (LMC) in ultraviolet light. The LMC, an active star-forming irregular galaxy, is in orbit about the Milky Way and is slowly being dragged down into the Milky Way. At some point in the distant future the LMC will eventually be disrupted and merged into it. This is thought to have happened to other small galaxies many times in the Milky Way's past – like all large galaxies, it is thought to be something of a cannibal. CREDIT: NASA.

16. Globular cluster 47 Tucanae. Globular clusters are among the most spectacular sights that it is possible to see with the eye through small telescopes. Most globular clusters are closer to the Galactic centre than the Sun and can be seen only from the southern hemisphere. However, there are a number which are easily visible from the northern hemisphere. Globular clusters lie in a spherical halo about the

Galaxy that stretches over 300,000 light years from the centre. CREDIT: Copyright Anglo-Australian Observatory. Photograph by David Malin.

17. Open cluster NGC 6520 and a dark nebula. All of the Milky Way's gas, dust and young stars lie in the middle of the disk of the Galaxy. Hence the youngest objects are almost always found to lie exactly on the Galactic equator (in the visible band of the Milky Way overhead). CREDIT: Copyright Anglo-Australian Observatory. Photograph by David Malin.

18. The Milky Way in infrared light. This image of the Milky Way was made by the COBE satellite and shows a disk of stars and dust very similar to that of NGC 4945 (plate 11). CREDIT: NASA Goddard Space Flight Center and the COBE Science Working Group.

19. The band of the Milky Way across the Galactic centre. This view of the Galactic centre is visible only from the southern hemisphere. It is a visible light picture of the central area of plate 18. Near the centre of this picture, but obscured by dust and gas, is a strong X-ray source: Sagittarius A. This source may be associated with a giant black hole in the centre of the Milky Way. CREDIT: Copyright Anglo-Australian Observatory. Photograph by David Malin.

20. The Cone Nebula. The nebula is 4,500 light years away in the southern constellation of Monoceros. CREDIT: Copyright Anglo-Australian Observatory. Photograph by David Malin.

21. The Tarantula Nebula. The nebula contains a young star cluster (that might well be a young globular cluster) known as NGC 2070. In the heart of NGC 2070 is R136. For many years R136 was thought to be a supermassive star with a mass more than a thousand times greater than the Sun. In the late 1980s, R136 was finally resolved as a tiny star cluster 1 light year across, containing hundreds of stars. Many of these stars will become supernovas, blowing the Tarantula Nebula away, leaving just the stars of NGC 2070. CREDIT: Copyright Anglo-Australian Observatory. Photograph by David Malin.

22. The Lagoon Nebula. The HST's WF/PC2 has revealed turbulence in the gas of the Lagoon Nebula. Again, the HST has been able to resolve details which would be impossible to see from the surface of the Earth due to the effects of the atmosphere. The different colours in this image correspond to light from different types of atoms: red from ionised sodium (sodium atoms that have lost some electrons), blue from ionised oxygen and green from ionised hydrogen. CREDIT: A. Caulet (Space Telescope European Coordinating Facility, ESA) and NASA.

23. A starburst in the galaxy NGC 253. The WF/PC2 took this picture of a starburst in the centre of the spiral galaxy NGC 253. The analysis of the starbirth in this galaxy shows that stars almost always seem to form in clusters, rather than individually. CREDIT: Carnegie Institution of Washington and NASA.

24. A giant region of starbirth in the galaxy NGC 604. This image was taken by the Hubble Space Telescope's WF/PC2 on 17 January 1995. Pictures were taken in many different colours. The changes in the images between different colours can tell us about the physics of the gas in which the young stars are embedded. CREDIT: Hui Yang (University of Illinois) and NASA.

25. Cometary globule CG4. The most spectacular picture of a cometary globule must be this of CG4, due to the chance alignment with the spiral galaxy in this picture. CG4 is rather faint, and the exposure time required to record it well has left the spiral galaxy slightly overexposed. CREDIT: Copyright Anglo-Australian Observatory. Photograph by David Malin.

26. Light echoes round Supernova 1987A. Since the explosion of Supernova 1987A it has become one of the most observed objects in the sky. The study of how the blast wave from the supernova propagates through space tells us a lot about the conditions and energy in the initial explosion. This picture was taken with the WF/PC2 in February 1994. CREDIT: Christopher Burrows (ESA/STScI) and NASA.

27. Supernova 1987A: before and after. It was only after Sanduleak −69° 202 exploded as Supernova 1987A that astronomers became interested in the star. This supernova was the closest since modern astronomy began and is visible to the naked eye (only in the southern hemisphere, however). The original star was thought to have been about twenty times more massive than the Sun and only a few million years old. CREDIT: Copyright Anglo-Australian Observatory. Photograph by David Malin.

28. The Crab Nebula. This image was taken on 5 November 1995 by the WF/PC2 on the Hubble Space Telescope. The image was taken in only a small band of colours corresponding to yellow light (the red colouring is arbitrary, but helps the human eye see some of the detailed ripples in the nebula). CREDIT: Jeff Hester and Paul Scowen (Arizona State University) and NASA.

29. The Vela and Puppis supernova remnants from ROSAT. These two supernova occurred so recently in astronomical history that the temperatures of the gas blown out of them are still incredibly high. Superheated gases such as these (known as plasmas) radiate very strongly in the X-ray band, making X-ray observations a good method of finding supernova remnants that may be invisible normally (unlike these, which are very obvious). CREDIT: Bernd Aschenbach, Max-Planck-Institut fuer Extraterrestrische Physik and NASA Goddard Space Flight Center.

30. A family portrait of Jupiter. This Voyager 1 picture taken in 1979 shows the massive planet Jupiter together with three of the four massive Galilean moons: Io, Europa and Callisto. Jupiter's Great Red Spot is also clearly visible in the planet's southern hemisphere. CREDIT: NASA Jet Propulsion Laboratory.

31. Jupiter's moon Io. When the Galileo spacecraft reached Jupiter it released a probe into the planet's upper atmosphere; the spacecraft then began an extended tour of a number of Jupiter's moons. Its first target in late June 1996 was Io. It revealed many changes in the surface of this moon since it was first studied in detail by the Voyager missions nearly two decades before. The changes are caused by the continuous and violent volcanic activity. CREDIT: NASA Jet Propulsion Laboratory.

32. Saturn from Voyager 1. This image of Saturn was taken by the Voyager 1 spacecraft just four days after its closest approach to the planet in November 1980. The spacecraft was looking back at the planet as it moved onwards past some of the Saturnian moons towards the outer Solar System. CREDIT: NASA Jet Propulsion Laboratory.

33. A global view of Mars from the Viking orbiter. This view of the Cerberus hemisphere of Mars is a montage of 104 images taken by the Viking 1 orbiter. The picture shows well what Mars would look like from a high orbit, lacking only a little haze around the edges which would be caused by the thin Martian atmosphere. CREDIT: NASA Jet Propulsion Laboratory.

34. A radar map of Venus. This view of the surface of Venus covers the entire northern hemisphere with the north pole at the centre. Most of the radar images that make this map come from the Magellan spacecraft, which mapped 98 per cent of Venus. The gaps have been filled with data from the Soviet Venera 15 mission and the Pioneer Venus Orbiter along with some information from the huge Arecibo radio telescope in Puerto Rico. CREDIT: NASA Jet Propulsion Laboratory.

35. Mercury from Mariner 10. The Mariner 10 mission has given us the most detailed information on the planet Mercury – a small, battered world. Mariner 10 made three fly-bys of Mercury and was able to map about one-third of the planet's surface, revealing a very cratered world. CREDIT: NASA Jet Propulsion Laboratory.

36. The Sun from the SOHO satellite. This image was taken in the extreme ultraviolet light of the Sun (light so energetic it is nearly X-rays) by the SOHO (SOlar and Heliospheric Observatory) satellite. This satellite is positioned 1.5 million km away, between the Sun and the Earth where the gravity of the two bodies cancel each other out (a Lagrangian point) and where it can observe the Sun continuously. CREDIT: NASA and ESA.

37. Halley's Comet from Giotto. The Giotto spacecraft passed close to the nucleus of this most famous of comets in mid-1986. For the first time a comet's nucleus could be imaged and the contents of its tail analysed from close quarters. CREDIT: NASA National Space Science Data Center and ESA.

38. Asteroid Ida and its moon Dactyl. The asteroid Ida orbits in the asteroid belt between Mars and Jupiter and was the 243rd asteroid to be discovered. Galileo passed within 11,000 km of Ida in August 1993 on its way to Jupiter, providing only the second close encounter of a spacecraft with an asteroid. CREDIT: NASA Jet Propulsion Laboratory.

39. The Earth–Moon system. This picture was taken by the Galileo spacecraft as it passed Earth for the second time in December 1992 before starting its three-year journey to Jupiter. It was taken from a distance of 6.2 million km (about sixteen times the Earth–Moon distance). CREDIT: NASA Jet Propulsion Laboratory.

40. The Moon in false colour. This image is made up of fifty-three pictures taken by Galileo on the same fly-by of Earth as for plate 39. The images were taken through three colour filters, which show how different areas of the Moon reflect different amounts of light of each colour. The different reflective properties provide clues as to the composition and age of these areas. CREDIT: NASA Jet Propulsion Laboratory.

41. The Earth from space. This picture again comes from the Galileo spacecraft which has provided some of the most stunning pictures of the planets since the Voyager missions. It was taken at about 14:10 GMT on 11 December 1990 on its first Earth fly-by. Galileo then passed through the inner Solar System before returning to Earth two years later for another flying visit. CREDIT: NASA Jet Propulsion Laboratory.

42. An impact crater in Namibia. This picture is another radar image, similar to that of Venus (plate 34), but this time of a region of Earth. This image was taken from the space shuttle's Spaceborne Imaging Radar, part of NASA's Mission to Planet Earth project. Satellite-based observations of the Earth are revealing much about the history of the Earth, and providing evidence of humankinds's huge influence upon the planet. CREDIT: NASA Jet Propulsion Laboratory.

INDEX

Page numbers in italic refer to illustrations